Fascinating Life Sciences

This interdisciplinary series brings together the most essential and captivating topics in the life sciences. They range from the plant sciences to zoology, from the microbiome to macrobiome, and from basic biology to biotechnology. The series not only highlights fascinating research; it also discusses major challenges associated with the life sciences and related disciplines and outlines future research directions. Individual volumes provide in-depth information, are richly illustrated with photographs, illustrations, and maps, and feature suggestions for further reading or glossaries where appropriate.

Interested researchers in all areas of the life sciences, as well as biology enthusiasts, will find the series' interdisciplinary focus and highly readable volumes especially appealing.

Carolyn M. King

Stoat in the Dock

Ecology and Management of Invasive Mustelids in New Zealand

Springer

Carolyn M. King
Te Aka Mātuatua School of Science
University of Waikato
Hamilton, New Zealand

ISSN 2509-6745 ISSN 2509-6753 (electronic)
Fascinating Life Sciences
ISBN 978-981-96-4561-9 ISBN 978-981-96-4562-6 (eBook)
https://doi.org/10.1007/978-981-96-4562-6

This Springer imprint is published by the registered company Springer Nature Singapore Pte Ltd.
The registered company address is: 152 Beach Road, #21-01/04 Gateway East, Singapore 189721, Singapore

Foreword

Humans have a long history of persecuting other predators—from bears and wolves to foxes and rats. The charge that stoats are responsible for there being so many endangered birds in New Zealand is simple, but the evidence is not.

It is not simple because many other pest mammals have also been introduced to this isolated and vulnerable archipelago. The kiore or Polynesian rat arrived centuries earlier and removed the most defenceless forest birds that had never seen a rat. It is still here, at least in small parts of the South Island. Others, like ship (roof) rats, Norway (brown) rats, brushtailed possums, ferrets, feral cats and hedgehogs are, like stoats, widespread, and also eat birds and their eggs and young. Feral pigs, several deer species and goats have changed forests by browsing and trampling. Forest area itself has been reduced by three quarters after two waves of colonisation, so that the original forest habitat itself is now gone or fragmented, especially in the accessible lowlands that are these days world-renowned for their green farms.

Dr Carolyn ('Kim') King has been a researcher, author, book and journal editor, historian, conservationist and lecturer in a science career spanning nearly six decades, and the globe. She is a world authority on stoats and weasels—both as pests (in New Zealand) and as valued natives (in her native England). In addition to 130 research papers, she has authored five books about mammal pests and their biology and control, and edited another eight, including three editions of *The Handbook of New Zealand Mammals*.

This book is a timely contribution from a true expert about one of the seven species (stoat, ferret, weasel, ship rat, Norway rat, kiore, possum) that New Zealanders aim to eradicate from the whole country by 2050.

One of my biologist maxims is that if you want to understand any species, you need to look back at where it came from. Who were its ancient enemies and what were its old foods? This book is a concise account of stoat biology and behaviour in both original and introduced ranges, explaining how and why stoats prey on birds, and how this beautiful and charismatic little animal became cast as a pest after its deliberate introduction to New Zealand. Part of that story is how forests themselves behave, and how episodes of mass seeding and fruiting influence the rat and mouse prey of stoats and thus the numbers and behaviours of stoats themselves.

In New Zealand we can now look back at 50 or so years of focused research and diverse attempts to limit stoat impacts. They have been eradicated from many islands and a few mainland ecosanctuaries, but reinvasion at all other trapping and poisoning sites remains inevitable without those few kilometres of water or a pest-fence as a physical barrier.

Read on. Here are stories from an old and a new land of humans and of stoats—both in their own ways enigmatic and impactful. Here is how nature, society and science are interacting to find the wild places—if any—still needing protection from stoats in the New Zealand of the future.

Senior Scientist, Manaaki Whenua—Landcare Research John Innes,
Hamilton, New Zealand
30 January 2025

Competing Interests

I received no external support for the preparation of this book. Sources of funding for my own research studies cited here are acknowledged in the relevant papers. I gratefully acknowledge formal permissions for all material not in my own copyright.

Sources of data and ideas in the published literature, and of key quotations, are given for each chapter, and in the captions to figures and tables.

Acknowledgements

During the > five decades of my research career, concentrated almost entirely on mustelids and their prey, I have met many of the people whose inspiring work is mentioned in the text here or listed in the chapter bibliographies, plus many, many others not listed. Mentioning some individuals risks offending others not mentioned, but I am hugely grateful to them all, and hope that they will all already know how much more I owe them than could be listed here. Therefore, I offer them this book in partial thanks for giving me the time of my life.

It is a truism, but still true after many repetitions, that authors of academic works never work alone. In the well-worn words of Isaac Newton, we can only stand on the shoulders of giants. Therefore, my institutional acknowledgements go with heartfelt gratitude to the many people and organisations with whom I have worked since first starting as a newbie graduate student in Charles Elton's Bureau of Animal Population in Oxford, a fortnight before he retired in October 1967 (Chap. 2). All the work described in these pages is based on studies begun then—my own and others'—and continued ever since I moved to New Zealand 4 years later. Grateful acknowledgements due to other institutions and their staff, past and present, include: the former Department of Scientific and Industrial Research Ecology Division and its Director John Gibb (Chap. 3); its descendant Landcare Research and many of its current and previous staff especially Andrew Veale, John Innes and Patrick Garvey; the former National Parks Authority and its Wellington officers Bing Lucas and Ben Thorpe (Chap. 4); the former Fiordland National Park Board and its Chief Ranger Alan Cragg (Chap. 5); the former New Zealand Forest Service and its staff Andy Leigh and John Gaukrodger (Chap. 6); the former HortResearch Bioengineering group and bioengineer the late Rod McDonald and his team (Sect. 9.5); Prof. Roger Powell of North Carolina State University (Chaps. 5 and 11); since 1987 and continuing, the Department of Conservation and many of its staff, especially Elaine Murphy and Craig Gillies; and since 1991 and continuing, the University of Waikato, librarians especially Cheryl Ward and Alistair Lamb, cartographer Max Oulton, administrators Fiona Martin and Klaus Reiter, technicians Sarah Campbell and Bruce Patty, current Vice Chancellor Professor Neil Quigley and successive Deans of Science especially Professors Warwick Silvester, Bruce Clarkson, Margaret Barbour and Chad Hewitt, plus the many graduate students whose work is cited here. Much more detailed acknowledgements are listed in all the source documents.

This particular book owes its origin to the questions I am often asked by academics interested in New Zealand but unfamiliar with the vast literature about invasive predators. During frequent talks to local conservation groups, I meet many people especially concerned to know more about the lives and potential management of stoats. Their curiosity stimulated this attempt to pull together the narrative presented here. New Zealand is one of the most significant known examples of the drastic impacts of introduced predators on endemic species and communities, still largely unknown to international audiences.

A wide-ranging account informed by a lifetime of working at the forefront of the field can provide a background helpful for updating the public debate. Such an update is timely because recent developments in predator control technology have led to enormously significant changes in conservation policies. Predator Free New Zealand PFNZ2050 has stimulated a recent surge of interest in the idea of evicting all invasive predators from New Zealand. Many widely held opinions about which are the worst of them, and what to do about them, circulate around the social media. Old ideas need to be updated, and some of the critically important assumptions that could derail the PFNZ2050 campaign need to be examined.

For indispensable and unpaid help during the long process of assembling this book, my most profound thanks go to Roger Powell and Andrew Veale, along with David Bolduc, Patrick Garvey, and John Innes. These indefatigable and knowledgeable colleagues made time to read a long MS with sufficient attention to make many useful comments and spot potentially embarrassing errors. Andrea Byrom, Craig Gillies, Dave Kelly, Elaine Murphy, Colin O'Donnell, Roger Pech, Roger Powell, James Russell, Dave Towns and Susan Walker checked the chapters concerning their own areas of expertise; and many, many others read separate sections and answered particular questions whenever asked. Over the much longer term, I remain indebted to the unflagging support of Mick Southern, Joe Miller and Ken Ayers throughout all my previous studies since my life-long fascination with mustelids began in Oxford in 1967. So if this book sounds rather like an extended personal journey, that's because it is.

All chapters comprise current reviews of each topic at the time of writing, formulated for readers without detailed experience of the history of mustelid research in New Zealand. The result ranges only lightly over this huge subject, but contains enough clues to lead to sources of further information, most now available online and many open-access. To assist overseas readers, a map of New Zealand common place names (Fig. 1.1) is offered as a general background (more detailed maps showing local place names in English and/or Māori are available on the internet). The text uses both the English and Māori names of animals, plus their scientific names in parentheses.

At Springer, Fumiko Yamaguchi and Kavitha Jayakumar provided constant encouragement and immediate support throughout the production process.

The cover photograph is by the late David Hallett.

Competing Interests The author has no competing interests to declare that are relevant to the content of this manuscript.

About the Book

This book is written for everyone who loves the natural world of New Zealand, and who works tirelessly in community groups and research institutions to repair and protect it. For them, I aim to offer some understanding of the history of New Zealand's long struggle against invasive mustelids, especially stoats (*Mustela erminea*). Unlike the commensal rodents, which could not have been prevented from disembarking from European and American ships independently, the mustelids were actually invited to come here. Analysing how and why this invitation was made, in all good faith and after over-ruling vigorous objections, may help those charged with making contemporary decisions to avoid repeating the mistakes of the past (Chap. 1).

The details of past work described here will certainly be out of date as soon as the book appears, but meanwhile it may be useful as an index to the vast but scattered stores of current information needed to interpret the present, and for planning new projects. With community activists, I grieve over lost biodiversity, but I also celebrate the range of new advances in technology, and the early successes of the Predator-Free New Zealand programme.

The stoat has long been recognised as a widespread and intelligent invasive predator in New Zealand, and a serious threat to native fauna, especially unprotected juvenile kiwi, a treasured national icon. Since the 1950s, National Parks rangers have worked to remove alien species from conservation lands, in the hope that native fauna were in fact benefitting.

By the 1970s it became clear that any attempt to control stoat populations artificially had to first know what controls them naturally. A previous study of weasels (*Mustela nivalis*) in UK (Chap. 2) provided the tools needed for comprehensive documentation of the biology of stoat populations in New Zealand national and forest parks (Chaps. 3, 4, 5 and 6). That background data has informed the history of the following long campaign against stoats, developed as part of contemporary conservation policies.

From the 1990s onwards, advances in technology permitted rapid progress in island eradications, safe and efficient use of new toxins, remote monitoring, advanced statistical analyses and genetics of wild stoat populations. All these are reviewed in Chaps. 7, 8, 9, 10, 11 and 12.

Chapter 13 describes how a new aspirational programme, Predator-Free New Zealand 2050, has galvanised community groups since 2016, with wonderful local results. Chapter 14 summarises the evidence against stoats, weasels and ferrets (*Mustela furo*) as contributors to the past and ongoing losses in New Zealand's biodiversity. Chapter 15 offers a review and some contemporary conclusions.

Contents

About the Author

Carolyn M. King I grew up on a Kentish farm, so have always been interested in the natural world. As a graduate student, I developed a strong and direct research interest in small introduced predators. I emigrated from the UK to New Zealand in 1971, straight from Oxford where I completed the first ever full-length study of the field ecology of weasels.

I worked with DSIR Ecology Division in Lower Hutt as a scientist specialising on introduced carnivores, especially stoats. In 1977 I left DSIR to concentrate on my family, but continued to write papers and books part-time. I have produced 130 papers in refereed scientific journals and book chapters, covering key aspects of the biology and management of weasels, stoats, ferrets, house mice, ship and Norway rats. My first book, *Immigrant Killers*, published by Oxford University Press in 1984, is still cited even though much of it is now out of date. In 2019 Palgrave Macmillan updated it with *Invasive Predators in New Zealand: Disaster on Four Small Paws*. My two best-known books, *The Natural History of Weasels and Stoats* (1989, 2007) and *The Handbook of New Zealand Mammals* (1990, 2005, 2021), both remained in print until they went into new editions.

From 1983 to 2009 I was a Scientific Editor for the Royal Society of New Zealand, as executive Editor of the *Journal of the Royal Society of New Zealand* until 2001 and of the *New Zealand Journal of Zoology* from 1991 to 2009, followed by 6 years as a Senior Editor for NZJZ.

In 1995 I joined the Department of Biological Sciences at Waikato University (Hamilton, New Zealand), and taught vertebrate zoology and ecology until retiring in July 2018. I am now an Emeritus Professor and a Fellow of the Royal Society of New Zealand, completely unfunded but continuing to speak and publish about stoats whenever invited.

List of Figures

List of Tables

Part I

Part I Setting the Scene

1 Introduction

Abstract

Stoats, weasels and ferrets were officially and deliberately imported into New Zealand during the 1880s. Influential but misguided pastoralists afflicted by plagues of European rabbits were convinced that the 'balance of nature' had been disturbed. They persuaded the Colonial Government to see predators as potential agents of biological control capable of restoring the balance. Over the decade 1883–1892, thousands of stoats and weasels were imported from UK, and tens of thousands of ferrets were locally bred, and liberated on damaged sheep pastures. Ferrets mostly stayed there, but stoats spread into forest. Weasels lacked their staple prey, the northern voles and lemmings, so are now rare.

Warnings from ornithologists of the risk to native birds were ignored, but stoats and ferrets have contributed to an ongoing massive failure in recruitment among unprotected populations of vulnerable native birds, including New Zealand's five iconic endemic species of kiwi. Meanwhile, both rabbits and mustelids have thrived. The results of this catastrophic, legal experiment in landscape-scale biological control proved the opposite of what was expected. With funding and material supplied by pest control agencies, New Zealand has turned out to be the best place in the world to study stoats, but for the saddest of reasons.

Keywords

Rabbits · Biological control · Kiwi · Unexpected consequences

C. M. King, *Stoat in the Dock*, Fascinating Life Sciences,
https://doi.org/10.1007/978-981-96-4562-6_1

1.1 Introduction

William Herbert Guthrie Smith was a pioneer philosopher, conservationist and sheep farmer, and the author of lyrical descriptions of the natural history and abundant native birdlife on Tūtira, his Hawkes Bay sheep station in the late nineteenth and early twentieth centuries. He was well ahead of his time in grieving over the vexed question of how best to protect native biota from invasive pests. The rapid destruction of native wildlife by introduced alien predators was certainly a tragic consequence of mismanagement by desperate and influential runholders faced with bankruptcy. Guthrie Smith saw it as much more than that—it was a matter of criminal irresponsibility.

> The freeing, firstly of rabbits, and secondly of weasels, in the Wairarapa was [a terrible mistake]….it was an outrage that any individual or local body should have been allowed to attempt to correct a blunder by a crime. (Guthrie-Smith, 1969: 354)

The isolated archipelago of New Zealand has been the unhappy recipient of all too many such invaders, so has become one of the world's hotspots for dealing with such unwelcome new arrivals. Protection of native biodiversity has become big business around the world. The foremost conservation organisation in New Zealand, The Royal Forest and Bird Society, has more than 80,000 members. For them, and for the public generally, the ongoing threats to our irreplaceable native biodiversity from invasive predators are a matter of great and enduring concern. Extensive and often publicly funded operations intended to control their numbers are urgent, and all too often conducted without reference to their important historical and ecological background.

Three European mustelids—stoats (*Mustela erminea*), weasels (*M. nivalis*) and ferrets (*M. furo*)—were, unlike rats, deliberately imported, with the support of the Colonial Government, as agents of biological control against rabbits (Sect. 1.2). Weasels look like smaller versions of stoats, and are even smaller than the rather different feral ferret. Comparing the way they have each adapted to such a radically different environment offers a unique glimpse of invasion ecology in a different world (King et al., 2017).

Guthrie Smith saw the consequences first hand, as the first cohorts of weasels over-ran Tūtira (Sect. 14.4). Mustelids failed as agents of biological control, but have succeeded all too well in adding to the heavy toll on native birds, lizards and giant invertebrates due to rats since the late 1200s (King, 2019). Both rats and mustelids have been among the key targets of organised predator control operations since the early 1970s. The future options available to such programmes will depend on what we know about which predators are responsible for at least some of the losses, and—more importantly—how their population dynamics work, sufficient to design and fund targeted management actions. Balancing the evidence for and against each suspect will be a process akin to a court case—hence my title, *Stoat in the Dock*.

The Problem Endemic kiwi once roamed throughout the forests, as described in lyric detail by early European explorers such as Ernst Dieffenbach (Bell, 1976) and Richard Henry (Hill & Hill, 1987). Stoats are aggressive, fearless, obviously carnivorous, and commonly active during the day. They are largely responsible for an ongoing massive failure in recruitment among unprotected kiwi populations, because kiwi chicks are very vulnerable to stoat predation. Hundreds of stoat traps are set every year by dedicated local community groups (Chap. 13). Many people mourn the loss of kiwi, and think of stoats as aggressive, vicious killers, and derive deep satisfaction from surveying a pile of dead stoats (Fraser, 2006). That reaction alone justifies a sharp focus on the problems of managing stoat populations (and of other mustelids). But intelligent pest control requires more than that.

Ideally, any sensible authority should always know what it is aiming for before starting a campaign, but when the risk is acute and the need to act is urgent, an immediate frontal assault is required. So the earliest efforts to tackle stoats in New Zealand relied on long-established traditional vermin trapping, based on decades of experience learned from English gamekeepers (Sect. 2.1). They knew how to remove stoats and weasels to protect nesting game birds, and that the numbers of mustelids varied from year to year, but not why they could never get the last one. They assumed that killing more stoats and weasels would be enough to protect their game birds, and, at least during the nesting season, they were often right (Tapper et al., 1996), but not for long.

The National Parks rangers who supplied the material for the first organised survey of stoat biology (Chap. 4) assumed that running a long roadside trap line, serviced from a vehicle in the course of other duties, would remove enough stoats to protect New Zealand's threatened native fauna. To the parks rangers supplying the necessary research material, the only interesting stoat was a dead stoat, and the only measure of success in helping them was an increase in the number they could kill.

But the basic assumptions of community ecology resist that simple conclusion. Surely, I said, attempts to control pest populations artificially could not be effective or justified until we first had some clue as to what controls their numbers naturally. Solving that problem was what my first job in the Orongorongo Valley project was supposed to be about (Sect. 3.1). Not everyone agreed with that logic, then or since.

Despite increasingly urgent efforts to get rid of them, weasels, stoats and ferrets all still survive in New Zealand, and all still contribute to the continuing decline in native fauna, especially birds and lizards. Moreover, there is a long list of other complicated and inter-related explanations for the historic decline in New Zealand's native species, which must be disentangled and analysed on their own terms. The mustelids are only part of a suite of introduced predators—especially rats and cats—so to know how to manage stoats we need to put them in context with all the other parts.

Since 2016, community groups and Department of Conservation staff have been galvanised by the Predator Free 2050 initiative (Chap. 13). It is a modern official, community-based and passionate campaign to eradicate seven key species of invasive predators from throughout New Zealand. To be most effective, those involved in predator control need to know their target species. Over the last few years I have attended many meetings of conservation groups discussing why stoats are so difficult to eradicate. Time and time again, they agree, we need to know more about stoats.

Outline Everything we see today has a history behind it, so it is much easier to understand what we are looking at now if we know how it came to be. So to examine this whole question from the beginning, we need to start with a bit of important background.

First, we need an account of how our knowledge of the ecology of stoats in New Zealand has developed over nearly six decades of research on them and on the ecosystems in which they live. Prior knowledge is essential to inform future decisions on when, where and how the damage due to stoats can be understood in context. The damage they do is well known, but was not and is not committed in isolation.

We begin with a historical survey that has value of its own, because we are now at a time when the pioneering generation of ecologists most active in New Zealand during the 30 years after 1970 is passing, and many of the new generation of scientists and pest controllers do not know the history of how or why stoats arrived in New Zealand, or the important research done on them before the advent of the internet. The rugged geography of New Zealand (Fig. 1.1) has provided its own restrictions on how the story has unfolded.

It might seem counterintuitive to start a book about stoats by talking at some length about weasels, but there is a reason for that. Most of the techniques that later proved so important to understand New Zealand stoats throughout the 1970s were worked out in advance in the late 1960s, when I was a graduate student struggling to understand English weasels. Both mustelid species have retained their fixed ancestral characters in their new home, although responding to their unfamiliar habitat differently. That meant that, by the time I moved to New Zealand in 1971, I was already armed with a toolbox of proven field methods that could be transferred to stoats mostly unchanged, and they gave me an invaluable head start—at least until new, stoat-specific techniques were needed. Later, I found that weasels are relatively rare in New Zealand, so it was helpful to have some useful species-specific background to refer to, summarised here in Part I.

Parts II and III are fully documented accounts of how our knowledge of the ecology of stoats in New Zealand has developed over the long processes of research on them and on the other mammals and birds with which they live (including, though to a lesser extent, their cousins, the ferrets and weasels). Part IV covers the current work driven by a new generation of stoat biologists armed with exciting new tools. New conservation initiatives like the Predator Free 2050 programme need the background information summarised here, so as to know what has been done before, what worked and what didn't.

The losses suffered by New Zealand's native biodiversity are well known, but were not and are due only to invasive predators. It has taken decades so far, but in Parts V and VI we discuss what evidence against stoats can be compiled by asking the birds themselves. Only by putting it all together can we begin to guess how,

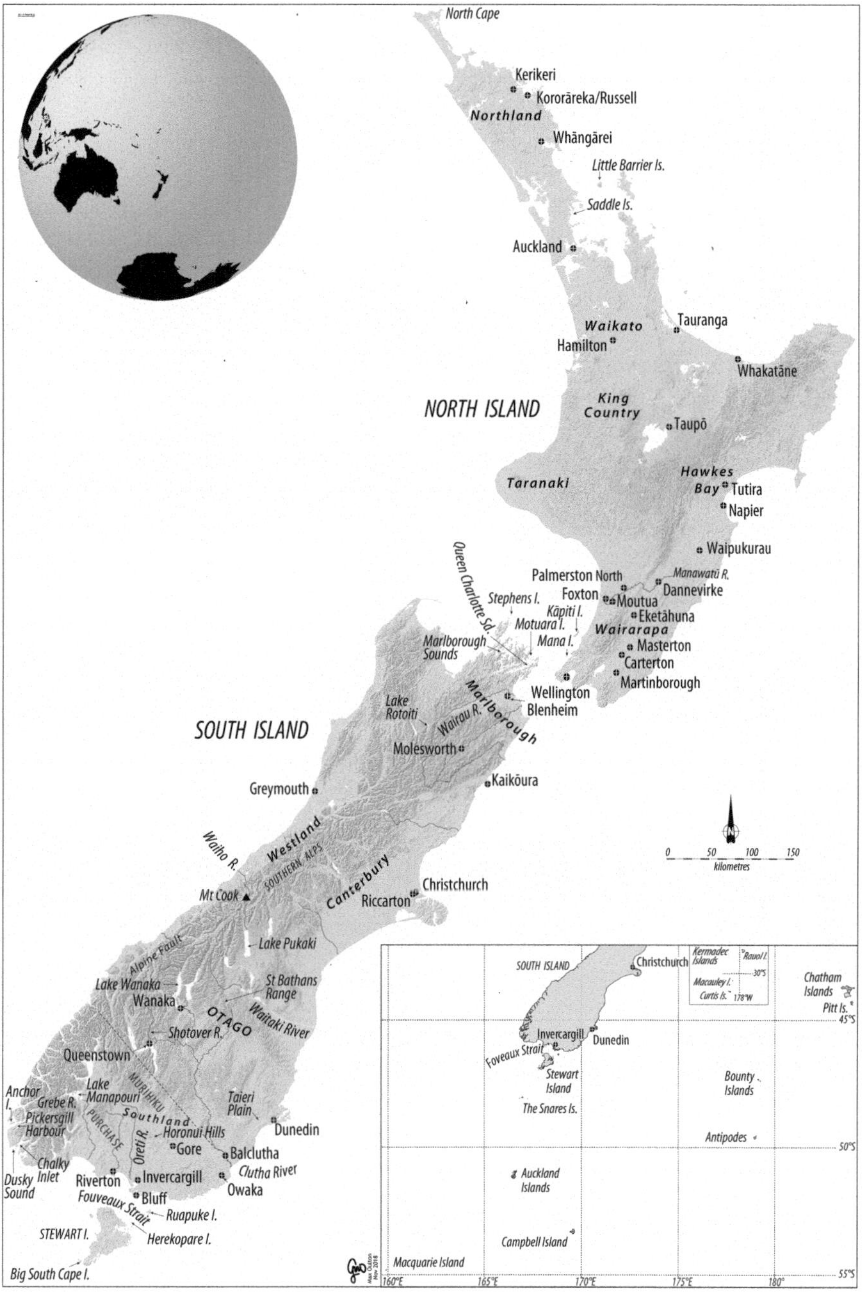

Fig. 1.1 Map of New Zealand and its southern offshore islands. (Cartography by Max Oulton)

when, where and how the damage due to stoats and ferrets might be minimised in the future. Much of what we have learned can be applied to invasive mustelids in other countries, such as American mink in Britain and Europe.

Rabbit, Stoat, Kiwi, Regret Stoats and ferrets are not the only European predators established in New Zealand. Feral cats have caused at least as much widespread damage to New Zealand's native biodiversity, and European rats much, much more, and both for much longer than stoats have done. So why do stoats cause so much more concern than rats? Conservationists usually overlook the far longer and wider destructive activities of the strictly nocturnal rats, but hate to see stoats running about during the day. The main reason is that stoats are very efficient at finding and killing kiwi chicks, so the 120+ years since the arrival of stoats have coincided with a drastic decline in numbers of kiwi, New Zealand's national icon.

New Zealand's long isolation has made it a classic example of the drastic impacts of human disturbance on endemic species and communities. Human New Zealanders are internationally known as Kiwis, so they tend to see any threat to the survival of kiwi birds as also threatening their own national identity. The first European voyagers landed only in 1769, but the settlers that followed brought multiple species of alien predators, including rabbits and stoats. The threat posed by stoats to the native kiwi surely counts as one of the most significant known examples of human mismanagement, requiring a massive conservation response.

This book is a history of stoat research for those still engaged in it, a resource manual for those seeking new remedies, and an analysis of a large-scale failed experiment on biological control for international audiences.

1.2 A Catastrophic Experiment in Biological Control, 1876–1892

The New Zealand archipelago and its native fauna had evolved throughout the Holocene in isolation from the rest of the world—that is, since before the evolution of the modern mammals. New Zealand's endemic biota was dominated by birds and lizards, but had not, as often claimed, always been free of all predators.

Native species were already well adapted over millennia to defend themselves from native predatory birds and reptiles. The mossy green plumage of the kākāpō, and the thick canopy overhanging kōkako nests, provide camouflage from native raptors hunting by sight during the day, but no defence against invading hunters with acute directional hearing. The nocturnal kiwi are too large to be much troubled by owls, but they have no defences against predatory mammals like rats, stoats and ferrets that hunt at night by scent. These and the many other non-native mammal species that invaded an island ecosystem like no other (King, 2019) are now spread throughout New Zealand.

Stoats are active both day and night, can climb and swim like natural athletes, can search out prey by scent and sound, and are flexible in diet and resilient in breeding strategy. They have turned out to be the very worst possible aliens to bring to a country formerly free of all ground-dwelling mammals, and, once established, they are extraordinarily difficult to eradicate. So why were they brought here in the first place? How much of the catastrophic decline of the once-abundant native biodiversity in New Zealand is due to them?

Historic Losses Actually, the slaughter started in the fourteenth century. Of an original estimated total of 242 known native birds (including 16 species of recent avian colonists, so not all 242 were present at the same time, but not counting several more recently extinct fossil species still awaiting description), at least 59 (24%) are now extinct (Tennyson, 2010). A large majority of the surviving endemic species are threatened to various degrees. So the transformation of the native biota was already well advanced, long before any stoats arrived.

The worst of the damage precipitated by the arrival of the first Europeans in the mid eighteenth century was due, among many other traumas, to the relentless nocturnal attentions of rats and feral cats accidentally introduced from ships since the late 1770s, plus widespread forest clearance. But to grieving conservationists, these were less obvious explanations than the *deliberate* importation of stoats (and ferrets, and weasels) during the 1880s (King, 2017c).

From the early 1870s onwards, New Zealand pastoral farmers had been fighting an unexpected but desperate battle against plagues of wild European rabbits. Domestic rabbits, an important source of trade articles and fresh meat, were normally carried on ships, but those that escaped seldom prospered. Trouble began when real wild rabbits were introduced in the 1850s, in all innocence, by settlers hungry for meat and shooting sport. The settlers were astounded by the unexpected consequences. Rabbits released on vast empty grasslands multiplied as rabbits do, to vast hordes competing with sheep for forage and undermining the pastures with burrows (Hodgkinson, 1876).

The Threat to the Colonial Wool Trade By the mid 1870s the damage caused by rabbits to sheep pasture was having a serious effect on the economy of pastoral farming. The only reason anyone could think of to explain why rabbits in New Zealand could reach numbers unknown in Britain was the absence of the rabbits' natural enemies. Too many rabbits meant not enough predators that eat rabbits. This was, to the afflicted runholders, surely a simple problem, with a simple and obvious solution. Australian experience ensured that foxes were already banned, so the task fell to the small mustelids—the stoats, weasels, and ferrets that were also clearly able to kill rabbits, but were not a potential threat to lambs. Pastoralists whose wool clips and profits were at stake pressured the Colonial Government to ignore any fears that the imported mustelids would add to the many serious pests of native birds already present.

A Catastrophic Experiment in Biological Control Alerted by friends in England, led by Professor Newton at Cambridge, ornithologists protested, and a Bill was introduced to Parliament in 1876 to prohibit the introduction of mustelids. It passed in the House of Representatives by a large majority, but in the Legislative Council (the upper chamber, since abolished) it was overruled by runholders whose economic self-interest outweighed any amount of sentimental or scientific interest in birds. It took until the turn of the century before they realized what a monumental mistake had been made.

Ferrets arrived first, and have generally remained in open country where rabbits are abundant, and native vegetation is damaged by grazing stock (King, 2017a). Ferrets now destroy thousands of ground-foraging birds and lizards every year (Sect. 14.3). But when ferrets were released in the cold high country in the hope they would halt the spreading grey blanket of thousands of invading rabbits, they could not meet the challenge, or survive the frosty winters. Some hardier predators were needed. Unlike ferrets, which were raised and distributed as domestic stock, stoats and weasels had evolved as specialist hunters of voles and lemmings in the cold tundras of the northern hemisphere.

So, in the early 1880s, under pressure from wealthy and desperate runholders, the Government began an official programme sponsoring the import of live stoats and weasels to New Zealand. Archives NZ preserves the critical letter from the Colonial Secretary commissioning one Walter Allbones, of Brigg in north Lincolnshire, to organise and transport shiploads of stoats and weasels round the world (Colonial Secretary, 1883). Many of the 17,000 or so gamekeepers then employed on the game estates of Britain were glad to take the opportunity to earn some extra pay. The little animals were easily caught in box traps, and purchased from gamekeepers for 6 shillings per stoat and 4 shillings per weasel. On the ships they were housed in small hutches, well cared for and accompanied by hundreds of cages containing pigeons to feed them on, plus hundreds of sacks of grain to feed the pigeons while they waited their turn for the chop.

The scheme ran until 1892, by which time thousands of weasels and stoats, and tens of thousands of ferrets, had been released alive on the rabbit-infested pastures of the new colony. In a detailed history of Tūtira, a classic New Zealand sheep station where he lived from 1882 to 1940, Herbert Guthrie Smith described the resolution, passed by a meeting of Hawkes Bay pastoralists with 'ferocious unanimity, that a guinea a head—and guineas were not then as now to be gathered from every manuka bush—should be given for every head of vermin dead or alive' (Guthrie-Smith, 1969: 354).

The Unexpected Consequences The importation of small predators was part of a huge, unplanned experiment in biological control, fortunately never since repeated on such a scale. Stoats certainly could and did kill rabbits in New Zealand as in Britain, but the pastoralists' expectation that they would stay on the sheep runs and hunt rabbits indefinitely was asking too much of them. Right next to the pastures were the forests, where birds, lizards and large endemic invertebrates nesting in trees were certainly not out of reach of agile stoats. And despite a hundred years' previous attention from Norway rats and feral cats, semi-wild dogs and human hunters, some of the native birds that foraged and nested on the ground still survived. Even without the runholders' misbegotten logic, the odds against these ancient lineages would still have been strong anyway, but the arrival of stoats greatly multiplied the risks.

Most of New Zealand's Acclimatisation Societies, and many individuals, foresaw and objected to the risk of harm to introduced game birds, but the government ignored them. In Britain, the comprehensive work of the Game Conservancy on the effects of predation on native partridges (Tapper et al., 1982, 1996) is an example of how much information is needed to understand predator–prey interactions within an animal community, and nothing like it was available for any New Zealand species until the late 1990s.

A detailed analysis of the transportation of stoats and weasels concluded that, against strong and repeated protests, at least 7838 stoats and weasels were imported from Britain in at least 25 shipments between 1883 and 1892 (King, 2017c). The total cost over the decade was probably at least NZ$1–2 million in today's money, with no effect on the numbers of rabbits. These historic details of the mustelid introductions cast an unconventional light on an old story, concerning how powerful economic interests and dominant cultural assumptions determine official policy, with implications for thinking about contemporary ecological globalisation.

The first official consignment of stoats was released onto pastures teeming with rabbits, with high hopes and the full protection of the law, on a peninsula in Otago's Lake Wanaka in 1884, where rabbits were extremely abundant. But within 6 years there were reports of stoats spreading over the main Divide of the Southern Alps into the forests of Fiordland and to other districts far from any known releases—including to places where there were as yet no rabbits—by the summer of 1889–90.

Both stoats and weasels were liberated only on rabbit-sick pastures throughout the North and South Islands (King, 2017b), but stoats can now be found almost anywhere, including in all the mainland national and forest parks. Weasels originally came in larger numbers than stoats, but the dominance of stoats and the absence of voles, normally the weasel's main prey, makes New Zealand an uncomfortable place for weasels. Ferrets mainly stayed on open country.

The tragedy of the Colonial Government's mustelid introduction policy is that there is in fact no evidence that mustelids can control rabbit populations in New Zealand, or indeed anywhere. Rabbit populations are primarily limited through bottom-up mechanisms (their numbers fall when their food resources and fertility rates are limited), rather than being top-down controlled by predators. Rabbits in most environments can breed more rapidly than predators can kill them, except when their numbers fall during temporary periods of food shortage, harsh and wet winters, and epidemics of disease, because their numbers are influenced by all these other things, but not by predation (Norbury & Jones, 2015).

Postscript There is one surprising consequence of this misbegotten nineteenth-century policy which no-one anticipated. Geneticist Andrew Veale discovered that the stoats transported to New Zealand have a higher genetic diversity (including at least five different haplotypes) than the single remaining haplotype now representing their ancestors in UK (Veale et al., 2015). This is the exact opposite of normal expectations, because the collection of a subset of animals to establish a new population cannot sample the full diversity of an ancestral population.

One likely explanation concerns the drastic crash in British stoat populations following the myxomatosis epidemic among rabbits in 1953–55. British stoats are rabbit specialists (McDonald et al., 2000), so the sudden loss of nearly all their major food supplies caused a massive crash in population numbers and in the genetic diversity of stoats in UK. By contrast, the animals collected from UK for export during the nineteenth century represent much of the former native genetic diversity of the species, all now lost except in New Zealand.

1.3 Early Observations on Stoats in New Zealand 1894–1964

Large forest birds were important traditional sources of food for Māori (Best, 1942). Then, for reasons unknown at the time, birds seemed to disappear even from previously reliable North Island hunting areas. For example, Te Heuheu Iwikau called a great hui at Pukawa in 1856, for which many forest birds were killed to provide a feast for visiting tribes. After that, the stocks of birds never seemed to recover. In evidence to the Waitangi Tribunal in 1993, Te Ahitahu Taiawhio stated that there had been no large-scale bird snaring since the war [1863–64], as the birds are much less numerous. The disappearance of birds from the North Island forests in the mid nineteenth century coincided with the arrival of ship rats after about 1850 (Atkinson, 1973), long before stoats appeared.

In the 1870s–80s, early European explorers normally set off on long surveying expeditions into remote inland South Island valleys carrying only dried foods such as tea, flour and sugar. They confidently depended on being able to catch enough of the abundant large birds kiwi (*Apteryx spp*), kākāpō (*Strigops habroptilus*), native pigeon/kererū (*Hemiphaga novaeseelandiae*), and forest parrot/kākā (*Nestor meridionalis*) to supply their need for fresh meat. When stoats reached Westland and Fiordland (now the Te Wahipounamu World Heritage Area, 26,000 km^2) in the 1890s, the last remnants of the ancient endemic fauna were still there. But by 1894, A. P. Harper was complaining that the strategy of explorers relying on birds for meat was no longer viable, because the birds were no longer a reliable source (Harper, 1896: 149–150). Richard Henry vividly described his desperate attempts to shift the survivors to safety on offshore islands, providing modern readers with a glimpse of our unimaginable loss (Henry et al., 2022).

Observers long familiar with the bird fauna of the South Island specifically blamed the arrival of stoats for the catastrophic losses they witnessed. Yes, it is true that stoats and ship rats assisted the final disappearance of the South I. subspecies of the bush wren (*Xenicus l. longipes*), NZ thrush (*Turnagra c. capensis*), laughing owl (*Sceloglaux a. albifacies*), saddleback (*Philesturnus c. carunculatus*) and kōkako (*Callaeas c. cinerea*), and aided the already advanced decline of the kākāpō, South I. takahē (*Porphyrio mantelli hochstetteri*) and little spotted kiwi (*Apteryx owenii*).

Most early writers thought the abundant birdlife they once observed represented the pristine state of the Westland/Fiordland fauna. But they never considered the undocumented previous declines of birds after the arrival of the Polynesian rat/kiore (*Rattus exulans*) in the thirteenth century, which is only now being understood (Gibbs, 2009; Worthy & Holdaway, 2001). Norway rats travelling with explorers and sealers on rat-infested ships in the late 1700s caused mass slaughter among seabird colonies around the Fiordland coasts (Sect. 14.2.9), a full century before stoats arrived. It is easy to blame stoats for damage done by other invaders.

1.3.1 Douglas 1894

Charles Douglas (1840–1916) was one of the tough breed of early explorers in the remote high country (Douglas, 1894: 71–75). He was an astute observer, and was among the first to distinguish the initial impact of Norway rats on forest birds from that of stoats. For the survivors of that previous rat-driven onslaught, the arrival of mustelids was the last straw, at least, in the southwest South Island. His reports, published in the *Appendices to the Journal of the House of Representatives*, are peppered with insightful predictions expressed in colourful language.

His report stated that ferrets and stoats moving west over the Alps had 'reached the Waiho Forks, and will soon be over the northern districts. They will finish the birds, and no doubt will get all the blame; but the process of extinction had commenced long before they arrived in the country. The Norwegian rat, which was no doubt responsible for some destruction, and which swarmed in the country at one time, is now becoming extinct from some cause or other. Perchance the ferrets will go the same way. But I fear that the kiwi and kakapo are doomed' (Douglas, 1894: 71).

1.3.2 Thomson 1922

G.M. Thomson (1948–1933) was the first author to attempt a detailed account of what was then known about the introduced animals of New Zealand. At that date, Thomson identified 48 species, of which 44 were brought in deliberately and four by accident, and he estimated that 25 species had already become truly feral. His attitude to the mustelids was uncompromising and widely quoted: 'Nothing in connection with the naturalisation of wild animals into New Zealand has caused so much heart-burning and controversy as the introduction of these bloodthirsty creatures' (Thomson, 1922: 70).

Thomson's section on the mustelids includes some illuminating first-hand accounts, although otherwise short on detail. But he concluded that they had not got rid of rabbits—indeed, that they do not even seem able to keep them in check.

1.3.3 Wodzicki 1950

Kazimierz Wodzicki (1900–1987) was the first Director of DSIR Animal Ecology Division, a Government-sponsored research organisation established in 1948 (Chap. 3). His first task was to gather the available information on past and present aspects of wild life problems in New Zealand, and so give a guide for future research (Wodzicki, 1950: 2). He organised the first survey of the distribution and economic importance of introduced wild mammals, by interviewing, and recruiting questionnaire responses from, more than 200 officers and field staff of other Government Departments, farmers, museums, Acclimatisation Societies and other observers.

The survey reported that the Department of Agriculture maintained ferret stud farms at least until about 1897, producing large numbers of young ferrets for release on pastures overrun by rabbits. Natural spread soon extended the distribution of ferrets around the country, and even into Fiordland in Richard Henry's time (Henry et al., 2022). There was some suggestion that the imported domestic ferrets included wild polecats, later disproven. Stoats and weasels were widely distributed, and similar to their European ancestors, but often confused with each other.

There were only two sources of numerical data available. One was supplied by fur dealer J.K. Mooney Ltd. of Dunedin. Over the 5 years 1944–48 inclusive, Mooney's warehouse handled 140,414 skins of ferrets and 50,957 stoats. The other came from the regional Acclimatisation Societies who kept records of the number of mustelid tails submitted for bounty. Over the decade 1939–1948 they paid out more than £3738 on 37,384 tails at two shillings a tail. Distinguishing weasels' plain brown tails from those of stoats with their black tail-tip is easy from material in the hand, confirming casual observations that, despite their early start, weasels were by then much scarcer than stoats. Despite the expenditure on bounties, Wodzicki cautiously concluded that the survey produced 'insufficient data to assess the effectiveness of these control measures'. Every later study of bounties agrees.

1.3.4 Marshall 1963

American Fulbright Research Scholar William (Bill) Marshall was invited to spend his sabbatical year in New Zealand, on leave from the University of Minnesota. He worked with Animal Ecology Division from September 1960 to June 1961. Bill and his assistant, Mike Fitzgerald, also used questionnaires to survey the abundance and distribution of all three mustelid species around the country, and mapped the responses they received from 272 correspondents (Marshall, 1963). As expected, stoats were seen more or less everywhere, but at that time ferrets had disappeared from Fiordland. Ferrets were still absent from large areas including Northland, much of Taranaki and eastern North Island, and the whole of Westland. The late arrival of ferrets in these districts helps to explain the better survival of kiwi in Northland, at least until the 1970s (Sects. 14.2.4 and 14.3.1).

From their survey correspondents, Bill and Mike collected 106 observations of stoats seen feeding, plus 52 carcases for gut analyses. This was the earliest attempt to document stoat diets, and although the samples were very small, they established the general picture that has been confirmed many times since (Sect. 4.4). Among potential stoat prey were birds, Australian brushtail possums (*Trichosurus vulpecula*), mice (*Mus musculus*), rats (*Rattus* sp.), lizards and hedgehogs (*Erinaceus europaeus*), fish and crayfish, rabbits (*Oryctolagus cuniculus*) (30 of 106 records) and birds (29 of 106).

1.3.5 Fitzgerald 1964

After Bill returned to Minnesota, Mike continued the work and expanded it into a systematic analysis of stoat biology for an MSc thesis. He ended up with 257 carcases, which provided data for the first-ever general analysis of the food habits of stoats (Fitzgerald, 1964). The same potential prey items turned up as before, and again in later collections using the same methods. Variations in proportions of species reported simply reflect the seasons and habitats sampled. Considering the undocumented large differences between the studies in the times and distribution sites of the early collections, this remarkable general similarly is probably a simple consequence of the small range of potential meals available to stoats in New Zealand.

Mike also conducted a much more meaningful pioneering study of stoat predation on ground nesting birds on Birdlings Flat, Canterbury. In ten stoat dens, he found scats and food scraps containing the remains of 227 birds, including 161 adult and 10 young skylarks/kairaka (*Alauda arvensis*), at least 58 banded dotterel/pohowera (*Charadrius bicinctus*) eggs, 82 mice, 16 lagomorphs and many skinks. The next survey using this valuable method was not done for another 40 years (Sect. 14.2.9). In due course, he joined DSIR Ecology Division, and made important contributions to the work of the Orongorongo Valley field station (Chap. 3).

1.4 The New Zealand Paradox

Stoats are widespread in the contemporary forests of New Zealand, fearlessly active during the day, well equipped with dagger teeth, and superbly adapted to the always risky life of a small hunter. Much has been learned here about these extraordinary animals, over decades of collaborative research effort.

The knowledge we have gained about the evolution and behaviour of mustelids generally has also been valuable to northern-hemisphere studies of stoats as native carnivores (King & Powell, 2007). The paradox is that, while our extensive collection of data is appreciated by those who want to increase the populations of small mustelids as valued natives, it has been driven by our need to understand how to reduce them as detested pests. Ironically, New Zealand has turned out to be the best place in the world to study stoats, but for the saddest of reasons. Funding and abundant samples of pest species are easy to obtain here.

This book attempts to describe that unfolding story. It may seem surprising to start an account of the biology of stoats in New Zealand by talking about English weasels. But as it turned out, my decades of trying to understand stoats got a head start from my previous 4-year study of their smaller relatives. The work of developing techniques for understanding weasels provided useful experience that could be quickly applied to later work on stoats, and much of what was learned from weasels in UK laid some important foundations for understanding stoats in New Zealand.

References

Atkinson, I. A. E. (1973). Spread of the ship rat (*Rattus r. rattus* L) in New Zealand. *Journal of the Royal Society of New Zealand, 3*, 457–472.
Bell, G. (1976). *Ernst Dieffenbach: Rebel and Humanist*. Dunmore Press.
Best, E. (1942). *Forest Lore of the Maori*. Dominion Museum.
Colonial Secretary. (1883). Letter 4 December to Agent General [re Allbones commission]. In *Outwards letterbook—Agent General*. Archives New Zealand R20557923.
Douglas, C. E. (1894). Report on the Westland Alps. *Appendices to the Journal of the House of Representatives, 1894*(C1), 71–74.
Fitzgerald, B. M. (1964). *Ecology of mustelids in New Zealand*. MSc thesis, University of Canterbury.
Fraser, A. (2006). *Public attitudes to pest control: A literature review* (DOC research and development series, 227). Science & Technical Publishing.
Gibbs, G. (2009). The end of an 80-million year experiment: A review of evidence describing the impact of introduced rodents on New Zealand's 'mammal-free' invertebrate fauna. *Biological Invasions, 11*, 1587–1593.
Guthrie-Smith, H. (1969). *Tutira. The story of a New Zealand Sheep Station* (4th ed.). A. H. & A. W. Reed.
Harper, A. P. (1896). *Pioneer work in the Alps of New Zealand*. T. Fisher & Unwin.
Henry, R. T., Hill, S., Hill, J., & Jaenecke, V. (2022). *Letters of a naturalist. The field accounts of Richard Henry on Resolution Island*. Putangi Publications.
Hill, S., & Hill, J. (1987). *Richard Henry of Resolution Island*. John McIndoe.
Hodgkinson, S. (1876). Report of the Rabbit Nuisance Committee. *Appendix to the Journal of the House of Representatives, 1876*, I–5.
King, C. M. (2017a). The chronology of a sad historical misjudgement: The introductions of rabbits and ferrets in nineteenth-century New Zealand. *International Review of Environmental History, 3*, 141–175. http://press-files.anu.edu.au/downloads/press/n2519/pdf/08_king.pdf
King, C. M. (2017b). Liberation and spread of stoats (*Mustela erminea*) and weasels (*M. nivalis*) in New Zealand, 1883-1920. *New Zealand Journal of Ecology, 41*, 163–177. https://doi.org/10.20417/nzjecol.41.29
King, C. M. (2017c). Pandora's Box down-under: Origins and numbers of mustelids transported to New Zealand for biological control of rabbits. *Biological Invasions, 19*, 1811–1823. https://doi.org/10.1007/s10530-017-1392-6
King, C. M. (2019). *Invasive predators in New Zealand: Disaster on four small paws*. Palgrave Macmillan. https://doi.org/10.1007/978-3-030-32138-3
King, C. M., & Powell, R. A. (2007). *The natural history of weasels and stoats: Ecology, behavior and management* (2nd ed.). Oxford University Press.
King, C. M., Norbury, G., & Veale, A. J. (2017). Small mustelids in New Zealand: Invasion ecology in a different world. In D. W. Macdonald, C. Newman, & L. A. Harrington (Eds.), *The biology and conservation of musteloids* (pp. 257–277). Oxford University Press.

McDonald, R. A. et al. (2000). The diet of stoats (*Mustela erminea*) and weasels (*Mustela nivalis*) in Great Britain. *Journal of Zoology (London), 252*, 363–371.

Marshall, W. H. (1963). *The ecology of mustelids in New Zealand* (DSIR information series 38). New Zealand Department of Scientific and Industrial Research.

Norbury, G., & Jones, C. (2015). Pests controlling pests: Does predator control lead to greater European rabbit abundance in Australasia? *Mammal Review, 45*, 79–87.

Tapper, S. C., Green, R. E., & Rands, M. R. W. (1982). Effects of mammalian predators on partridge populations. *Mammal Review, 12*, 159–167.

Tapper, S. C., Potts, G. R., & Brockless, M. H. (1996). The effects of an experimental reduction in predation pressure on the breeding success and population density of grey partridges *Perdix perdix*. *Journal of Applied Ecology, 33*, 965–978.

Tennyson, A. J. D. (2010). The origin and history of New Zealand's terrestrial vertebrates. *New Zealand Journal of Ecology, 34*, 6–27.

Thomson, G. M. (1922). *The naturalisation of animals & plants in New Zealand*. Cambridge University Press.

Veale, A. J., Holland, O. J., McDonald, R. A., Clout, M. N., & Gleeson, D. (2015). An invasive non-native mammal population conserves genetic diversity lost from its native range. *Molecular Ecology, 24*, 2156–2163. https://doi.org/10.1111/mec.13102

Wodzicki, K. A. (1950). *Introduced mammals of New Zealand: An ecological and economic survey*. DSIR Bulletin 98.

Worthy, T. H., & Holdaway, R. N. (2001). *The lost world of the moa: Prehistoric life of New Zealand*. Indiana University Press.

The Weasels of Wytham Woods, 1967–71

2

Abstract

Most of the simplest techniques needed for studying invasive stoats (*Mustela erminea*) in New Zealand were developed, all unknowing, during a preceding doctorate study of weasels (*M. nivalis*) in England, the native country of both species. Both retained their hard-wired genomes and natural behaviour during and ever since the transfer, so the procedures developed and problems solved in the 1960s turned out to be applicable (with a few modifications) to work on stoats throughout the 1970s and 80s. These included the following. Twice daily live trap inspections, 1 week in three; the best trap lure (rabbit scent renewed daily); white mouse bait to provide food for a captive, easily distinguished from natural prey in the scats; individual identification by ear-tags and belly patterns; field anaesthesia (required for handling and tagging); gut analyses; home range size and turnover; estimated impact on woodland rodents; incidence and cure of raiding nest boxes of titmice (Paridae); collecting large samples of carcases from professional trappers; routine carcase analyses; skull cleaning; age determination; life tables; mortality and survival rates; longevity; and pre-computer data analyses using edge-punched cards and simple non-parametric statistics. These data also help understand weasels in New Zealand which are too scarce to provide local information.

Keywords

Live trapping weasels · Field anaesthesia · Age determination from carcases · First population analyses

C. M. King, *Stoat in the Dock*, Fascinating Life Sciences,
https://doi.org/10.1007/978-981-96-4562-6_2

The Bureau of Animal Population (BAP) was established in Oxford by Charles Elton, the pioneer of scientific animal ecology, and it was universally known and respected as Elton's BAP (Crowcroft, 1991). I joined it as a new graduate student 2 weeks before Elton retired. Later the BAP became the Animal Ecology Research Group, part of Oxford University's Department of Zoology.

The University owned a private estate called Wytham Woods, four miles outside the city. Most members of the BAP, and the neighbouring Edward Grey Institute of Ornithology (EGI), had conducted their field studies there since about 1947. Mutual support and swapping of data between students and staff working on different aspects of the same ecosystem made it a great advantage to work in such a place (Savill et al., 2011). Weasels are difficult to work on anywhere, but in such a supportive environment I survived learning some important lessons the hard way. It is worth spelling them out here because they became very useful later (Chap. 4).

2.1 First Catch Your Weasels

Marley Wood (26.8 ha) was part of the Wytham estate, a semi-isolated patch of mixed woodland surrounded on all but half a side by fields or plantations. My supervisor H.N. (Mick) Southern and I decided to set up my field study on the weasels that lived there. The first requirement was to obtain a stock of large, wooden box weasel traps from Whitlock Bros Ltd., a company that produced them for supply to game keepers.

Live Traps The traps were constructed as a simple wooden tunnel fitted with a see-saw treadle, with glass panel at the far end supposedly increasing the chances of enticing a weasel into an innocent-looking run-through passageway. The treadle was counter-weighted to exclude animals smaller than female weasels (under 35 g), and a hole was drilled in the side to allow a pin to lock the treadle open between trapping sessions. There was space for bedding between the inside end of the treadle and the glass.

Open woodland offers no predictable runways, so a weasel could range widely without finding a trap. Traps set across a broad forest floor have the problem of luring weasels into the right vicinity and then engaging their curiosity. The trap line was 2.5 miles round, and had to be inspected twice a day. In the first 6 months, 72 trap rounds in all, produced only seven captures of five weasels, four once only and one three times.

The project was in danger of being abandoned, until Den Wood, the field officer in charge of protecting the neighbours' lambs from foxes, told me an old gamekeeper's trick for encouraging weasels to enter the trap. It required the fresh gut of a rabbit or hare (which he supplied for me), tipped into a container, and the scent of it stirred around with a stick and then wiped on the entrance to the trap and on rocks and stones leading up to it. That changed the results, just in time (King, 1973). Assuming that weasels would be more active at night than during the day, I renewed the rabbit scent on the traps every evening. In fact it took twice as long for all the traps to be set before one of them caught a weasel at night (10.4 h) compared with during daylight (5.4 h). The same conclusion was confirmed later from radio-tracking (Sect. 9.2).

Captured weasels were marked with small metal ear-tags, inserted flat (under anaesthetic) into the tiny strip of cartilage at the base of the ear. A recaptured weasel with characteristic tear in an ear showed where a tag had been pulled out. In most other species, the identity of that animal might then be lost. Fortunately, British weasels offer a backup code. In their subspecies (*Mustela nivalis vulgaris*, but not their northern relatives, *M.n. nivalis*), the line between the brown fur on the back and the white of the belly is irregular. The white area always includes a brown spot behind the angle of the mouth, and often also other odd spots or large patches of brown on the belly. The details of the belly pattern are individually unique (Linn & Day, 1966), and constant through successive moults. A quick sketch of the belly pattern on the back of the capture record card for every new weasel ensured that even those individuals who lost their ear-tag could always be identified.

Handling Trapped weasels cannot be handled safely while conscious, so every live-captured animal was transferred to an anaesthetising box with a sliding metal door, a glass front, and a small hole in the top. When the glass ends of the closed trap and the box were held tightly together and the glasses removed, it was possible to gently persuade the weasel to move from the trap into the box (Fig. 2.1).

At that time, anaesthetic ether could be bubbled quite safely into the closed box through a simple tube. There was no standard dose—only a series of predictable

Fig. 2.1 Handling an unconscious weasel in the field. Live trap at top; ether reservoir, bubble tubes and small anaesthetising box at lower left. (Photo by Author)

responses from which it became easy to judge, by watching the weasel's reaction, when to stop. Modern Health and Safety officers would be appalled (Gorini et al., 2013), but no-one then thought there was any danger in using anaesthetic ether in the open air, and the weasels agreed. Of 320 anaesthetisations I performed in the field plus a further 50-odd in the lab, all except one recovered with no sign of any ill effects. (Anaesthetic ether is no longer available. New regulations require use of isoflurane and air in the lab or via a portable battery-powered unit).

Once unconscious, the opportunity to get through a list of tasks was rather short. They included: reading or renewing the tag, checking or redrawing the belly pattern, taking a weight (by suspending the unconscious weasel by the scruff of the neck from the crocodile clip on the end of a spring balance; see Fig. 3.1c), examining the canine teeth, collecting scats and the falling fleas (also anaesthetised), and estimating reproductive condition. Occasionally another quick visit to the anaesthetising box might be needed.

As well as the rabbit gut lure to encourage entry, weasels also needed to be fed whilst trapped. A dead white mouse provided an acceptable meal in a waterproof package. Providing some food was not only important to the survival of the trapped weasels, but also pushed the remains of its last meal through the gut, increasing my chances of collecting a useful scat. The fur of the white mouse was easily distinguished from any natural food, which helped to make the diet analysis easy.

Collecting location records for female weasels is especially difficult (Erlinge, 1974; Lockie, 1966). On the other hand, weasels caught in live traps set for rodents are much more often females than males (King, 1975). The difference is not due to any attribute of the traps, but to the fact that traps set for small rodents are always laid out closer together than kill traps set for weasels by gamekeepers. Additional reasons for this bias are that females travel less each day and have smaller home ranges that enclose a smaller number of traps laid out at any given density (Erlinge, 1974).

By the end of the study, I had recorded 336 captures of 25 weasels from September 1968 to June 1970, of which 16 were resident in Marley Wood, and nine others passed through without staying for long. The capture records were mapped at the end of each week's trapping, and eventually the aggregated records allowed rough estimates of home range boundaries. Seven resident males (but never more than four at one time) held home ranges of 7–15 ha. Neighbouring home ranges apparently overlapped, in a sort of 'communication zone' where each left scent messages for the other, but so far as the trapping records could tell, they avoided actually meeting each other. The females were, as expected, extremely hard to catch, but from the few captures recorded it seems that they lived in disproportionately smaller areas of 1–5 ha.

2.2 Weasels Living in a Woodland Community

Weasel scats contain only hairs, feathers, and bone fragments, which can be distinguished by a combination of minute differences in their structure. Hairs have a surface array of scales or chevrons that can be revealed by embedding them in warm gelatine. Stripping off the hair when the gelatine is dry can show the impress of surface patterns. The hairs of mice and voles also have different shapes in cross section, which are essential to distinguish otherwise similar species.

Diet The list of potential food items to be identified is relatively short , since weasels have evolved as specialist predators of small vertebrates (rodents, rabbits, birds, and lizards). In their native habitats they eat insects, worms, or vegetable matter such as berries only when extremely hungry. Moreover, since it is fair to assume that weasels never eat hair or feathers except in the course of eating the animal to which they were attached, it is possible to make a rough estimate of the total number of individual prey eaten. Because metabolism in weasels is so rapid, stomachs contain undigested meat only from very recent meals.

With that information, the diets of weasels could be deduced by standard methods. The procedures are straightforward, and the quantities involved are small, dry and inoffensive to the nose. Altogether, 344 samples (scats) were available from weasels detained in the live-traps or visiting locked traps between sessions. In the period between September 1968 and March 1970, 215 of these samples containing wild-caught prey were deposited by known individual weasels whose home ranges and movements were already mapped. A series of samples from a known home range over time made it easy to check for any relationships between individuals in diet and habitat and season (King, 1980c).

Deciding what the figures mean is far more difficult, because it is not always easy to derive useful information from food remains. To begin with, simple lists of prey eaten by weasels from different places will be biased toward the season, the habitat, and the ages and sex of the weasels most often represented. Likewise, samples to be compared must be matched for origin (from weasels of similar sexes or ages), or caught in the same season, and larger samples give the most reliable information. Any analysis has to reach some compromise between splitting, to avoid compounding different effects in one sample, and lumping, to increase the number of samples.

The Wytham weasels ate the available small rodents more or less in the order of their abundance. Bank voles were the most common of the small rodents in the wood, followed by wood mice, both averaging about 21 g each, smaller than the chunkier field voles (averaging 25 g) that lived in grassland habitats. Few field voles lived in the wood, but some were taken by weasels whose home ranges extended to a neighbouring young plantation where thick, ungrazed grass provided the field vole's favourite habitat. And yes, the prey species taken by the six individual male weasels, who supplied 60% of the 245 items identified, did indeed reflect differences in the habitats of their home ranges and the seasons during which they contributed scats. Wytham had very few rabbits and no rats or stoats at the time. There were plenty of shrews and moles available, but the resident weasels ignored them both.

The conclusion suggested that the poetic picture of a weasel, as "the Nemesis of Nature's little people," is rather different from the real one. In a face-to-face encounter, the odds are usually in favour of the weasel, but the weasel has first to make the encounter. In this, the long-term odds are certainly in favour of the vole. Besides, the meeting of a single vole and a single weasel conveys nothing about the impact of weasels on the numbers of voles, since the fates of populations cannot be judged from the fates of individuals.

Predatory Impact One of the main reasons why predation by weasels, or any other predator, is such a tricky subject to investigate is that it demands simultaneous documentation of the numbers of weasels and their prey. That is too much for a single student to do in a limited time, unless that student worked in Wytham.

John Flowerdew, fellow-student also working under Mick Southern, generously shared his estimates of the numbers of bank voles and wood mice available in Marley Wood during the year (September 1968–October 1969) when our field records overlapped. At that time, the combined density of these little rodents over the whole study area ranged from 21 to 39 individuals per hectare per month, at a time when a male weasel's home range could be up to 15 ha. Together we estimated that the resident weasels removed an average of 8% of the bank vole population per month, and 10% of available wood mice.

But this rate of loss varied hugely, and apparently accounted for only a small proportion of the 12–55% of voles and mice disappearing each month, and for perhaps 14% of the total production of rodents in the forest (Hayward & Phillipson, 1969). Some weasels learned that free food was a price worth paying for temporary inconvenience, and occasionally one of the regulars, released from a trap at the start of the round, would make a quick short cut across its territory to get another free meal in a later trap. Even so, the resident weasels were always hungry. Three of the longest-resident males died after severe loss of weight.

Predation by weasels also had little effect on the average survival time of individually marked mice. For example, in September 1968, 55% of bank voles known to be alive on John's control grid disappeared, whereas only 4% of the total population estimated to be present in the whole wood were eaten by weasels. Of course these estimates were very rough, and sampled only a small area of Marley Wood, but predation by weasels would have had to be impossibly greater to have had an observable effect on the numbers or survival of small rodents in the wood, especially in the summer of 1969 when there were rather few weasels in residence. Naturally, these estimates also had to rest on a series of assumptions, but it was reassuring to find that the months when the weasels' toll on rodents was high coincided with the months when most weasels were present.

On the other hand, weasels were not the only important predators resident in the wood. Tawny owls (*Strix aluco*) also relied on small woodland rodents for more than half their diets, and competition between owls and weasels could not be entirely avoided, because their common resource was not renewed rapidly enough. Mick Southern's long-term study of tawny owls showed that they were long-lived, they held stable territories despite the ups and downs of rodent numbers, and they hunted only at night (Southern & Lowe, 1968).

When food was scarce, the owls sat tight on their territories and deferred breeding until better times. The Marley Wood weasels lived frantic breed-fast-die-young lives, and they could reach rodents inaccessible to owls, down tunnels and under thick cover. They hunted mostly during the day—which I thought at the time was largely because they were also hunted by the owls, but radio tracking showed that they usually are diurnal in any habitat (Sect. 9.2).

Mick's estimates of the prey of tawny owls in another part of Wytham found that the combined density of wood mice and bank voles remained between roughly 10 and 30 rodents/ha throughout the 1960s and 1970s. He reckoned that his tawny owls removed on average 15–33% of the mice and voles each 2 months. This regular toll taken by the permanently resident tawny owls was much more constant, and at times it was substantially heavier, than that taken by the weasels (Southern & Lowe, 1982).

Wytham Wood has also been the site of more than 50 years of research on the population dynamics of tits (Paridae). These small birds nest in natural holes in trees, but also readily use the 1000+ artificial wooden nest boxes provided by researchers from the Edward Grey Institute (EGI). Their annual records of nesting success included data on which nests were destroyed by predators, most often by weasels.

For the first 10 years of the EGI's monitoring, until 1957, weasels raided only 0–8% of the nest boxes per season. Then, suddenly, up to half of the boxes were raided in each of the next few years. It turned out that the highest number of boxes raided in any 1 year coincided with seasons when small rodents in the wood were fewer than usual. Since 1976 the installation of concrete, weasel-proof nest-boxes has deterred the weasels from visiting the eggs and chicks, and reduced the rate of nest predation from 30% to less than 5% a year (McCleery et al., 1996). The weasels no doubt continue to raid the unprotected nests of other birds, as they do all over the world.

2.3 Weasel Populations

The burning question among animal ecologists in the late 1960s was: What controls natural populations of animals? Why do the well studied English garden birds—and the tawny owls in Wytham—seem to maintain roughly constant numbers from year to year, while field voles and lemmings seem to go through semi-regular cycles, and Australian house mice are capable of spectacular irruptions?

Sampling In an ungrazed meadow, there might be from hundreds to thousands of grass plants for every vole, and from scores to hundreds of voles for every weasel. On the other hand, because weasels are the smallest of the warm-blooded predators, they can be much more common than the foxes, feral cats, hawks, and owls that also eat voles, even though they can readily fall victim to any of these.

One might expect from this that it should be possible to collect large samples of weasels, and that their population dynamics would be among the best studied of all carnivores, but that is not so. The vernacular name of the smallest of the weasel family resident in Britain was and is the 'common weasel', but those living in Wytham were in fact not common at all.

Weasels are too small and secretive to be observed directly (except in occasional lucky glimpses), so collecting a large enough sample to study how their populations work has to be done indirectly, by routine large-scale kill-trapping. Suitable field techniques were worked out in the nineteenth century by fur trappers, naturalists, and gamekeepers, but were not applied to scientific studies until the 1960s.

Worse, weasels are not evenly distributed in all habitats, and even where they do live, they may be abundant 1 year and scarce the next. These variations make it hard to collect a statistically respectable sample from such a slippery, moving target. So, even after accepting some compromises with ideal methodology, only first-order approximations are possible. But at least one can aim to be mostly accurate if not precise, rather than precise and wildly inaccurate.

Changes in the behaviour of weasels in response to bait, and to variations in natural food supplies are serious, but not necessarily fatal. On the one hand, although baited traps catch weasels more often than unbaited ones, both do show the seasonal variation in capture rate that normally reflects a real variation in density. On the other hand, small mustelids in general maintain smaller home ranges and hunt over shorter distances when food is abundant than when it is scarce, so there is usually some variation in capture rate with prey abundance. Even so, in the 1960s when no other methods could give any clue as to how and where weasels lived, kill-trap results with all their faults could still be useful for studying weasels at landscape scale.

The "vermin books" of well-organized traditional game estates used to contain long sets of figures showing the numbers of predators killed every year. Weasels and stoats were always among the greatest enemies of the old-style gamekeeper, for whom every kill was a source of satisfaction. Some keepers used to hang the carcasses along a strategically located fence, accumulating a macabre "gibbet"—a display intended to demonstrate to their employers the good results of their hard work. So keepers have long known that the relative numbers of weasels caught could show massive swings in from year to year. Most keepers are interested to learn more about weasel populations by collecting carcasses of dead weasels and passing them on for analysis.

This ready-made source of material became an ideal second-string study, a reassurance policy in case of failing to get enough respectable results from livetrapping. Carcases of weasels were supplied by three game estates representing southern and northern England and southern Scotland. A standardised routine for processing the carcases included: recording weight and measurements, removing the skin to examine the moult stage from the inside, plus the guts and reproductive organs for separate analyses, and the skull and baculum for cleaning (King, 1980b). In the event both live trapping and carcass analyses studies paid off.

But this valuable material could not be used to construct a first preliminary estimate of the population dynamics of these tiny carnivores until the carcases could be classified by age. The biggest problem was that it was impossible to do that without reference to known-age specimens, of which I had none. A solution arrived from an entirely unexpected source.

2.4 Weasel Birth Certificates

In Branschweig, Germany, Fritz Frank had been maintaining a breeding colony of captive weasels for his own interests in reproductive biology. His animals were kept in semi-natural conditions and fed only on live field voles, which were bred in an adjacent facility. Up to 94 weasels passed through his hands over the years, of which he cleaned the skulls of 44 that died (26 males, 18 females), and the bacula (penis bone) of 16 males. He kept the skulls, complete with their dates of birth and death, in his archives.

Known Aged Reference Material Fritz had the only known collection of known-aged skulls closely representing those of wild populations. In mid July of 1969, he offered to share this unique and absolutely vital resource with me. He would not, naturally, allow me to take them back to Oxford, but he did offer to host me for a week if I chose to return to his lab and study them there. So in February 1970, I did that, and the photographs and measurements I took then became the foundation of my own method of age determination. Frank's generosity constituted a decisive moment for the study of weasel population dynamics.

The skulls of the British weasels supplied by gamekeepers could be readily cleaned by dermestid beetle larvae. Their tiny jaws stripped the meat out of every aperture, leaving only perfectly clean bone, and a few larval skins. A close analysis of both German and British collections allowed a summary description of five stages in the morphological changes detectable in weasel skulls as they grew from newly independent infants to old adults. When allied with a known date of death, and (in males) the weight of the baculum, I found that changes in the shape of the skull could be used to estimate the ages of wild-caught weasels. It became known as the date-skull-baculum (DSB) method (King, 1980a).

The most important feature to observe is the postorbital constriction, a gradual, measurable narrowing of the skull behind the eyes, expressed as the ratio between the inter-orbital and post-orbital widths (Fig. 2.2). Museums often hold substantial collections of weasel skulls, so this method is simple, requires no special equipment, and is most useful for distinguishing young animals, not yet full grown, from among a set of skulls and bacula that must not be damaged. The required measurements can also be taken in the field, where full skull cleaning equipment is not available, simply by peeling back the temporalis muscle from the bone on the top of the head. The very oldest individuals might also show exposed dentine on the carnassial teeth.

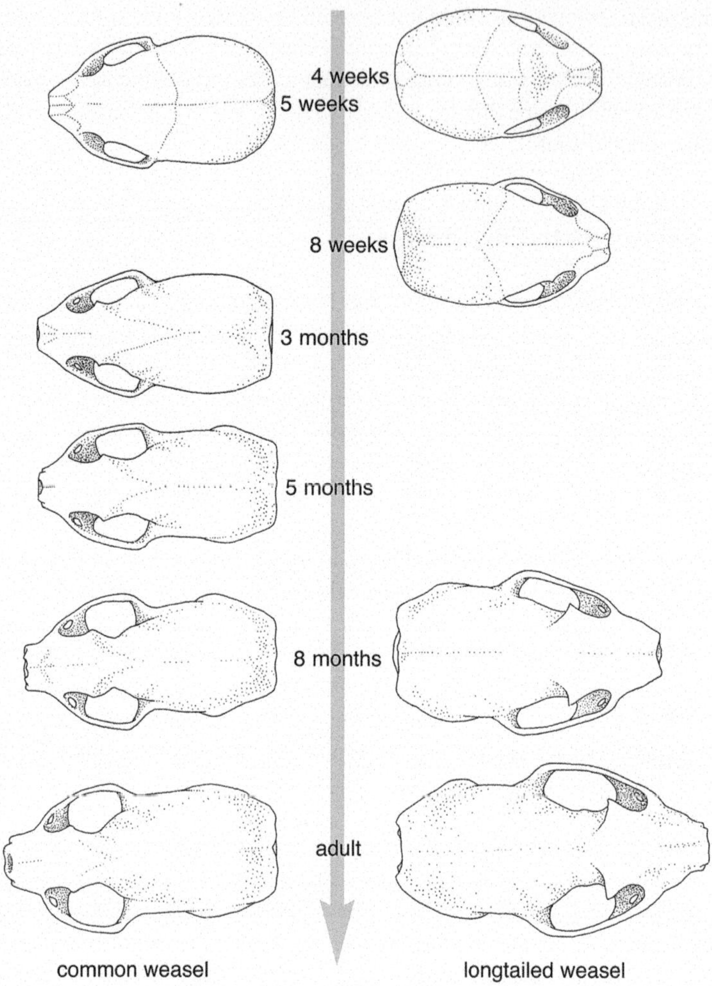

Fig. 2.2 (Left) The gradual narrowing of the postorbital constriction in known-aged *Mustela nivalis* bred in Germany. (Right) Known-aged *Neogale frenata* bred in New York (Hamilton, 1933), which are larger than *M. nivalis* but not as large as are New Zealand *M. erminea*, showing the same process but the greater development of muscle attachments and zygomatic arches in larger weasels. (Drawing by Consie Powell)

The development of the postorbital constriction, allied with features of the baculum and other anatomical features known to be related to age, allowed classification of wild-caught common weasels into two, or at most three, year classes. It proved quite reliable for the first 6 months after the young begin to appear in the population, though less so in late winter and spring. It can also be applied to stoats and to longtail weasels (*Neogale frenata*), but with different definitions of age classes. For equivalent data on known-aged New Zealand *M. erminea*, see Sect. 5.1.

With a method of estimating the age at which a weasel died it became possible, for the first time, to make some first guesses on how long they lived, and what effect trapping might have on their numbers. The disadvantage of the DSB method is that it cannot distinguish between the year classes of full-grown animals. This does not matter much for weasels, because very few of them live for as much as 2 years, but if histological machinery is available and destructive processing is permitted, the more complicated but more reliable method of counting cementum layers in the teeth is better.

That method was later tested by Robbie McDonald with another, larger sample of British stoats and weasels collected from gamekeepers (McDonald & Harris, 2002). He showed that the canine teeth of both species do have layers in the cementum (Sect. 11.4). At that time, there were no known-age standard tooth sections to confirm that the layers are annual, but it is probably safe to assume that they were.

2.5 How Long Do Weasels Live?

Population Analyses Unlike the stoat and many other members of the family Mustelidae, weasels implant their fertilised ova into the uterus soon after mating, and proceed straight into an active pregnancy averaging six embryos of equal sex ratio. In normal years for rodents, adult females produce one litter in spring, and their young do not breed until the next season. But when food is super-abundant, well-fed adults can produce a second litter of six, and the early-born females can breed in the year of their birth, thereby adding (in theory) three more sets of six.

That means that, during a very good year for rodents, a huge annual cohort of young weasels can be added to the population, at what amounts to compound interest. The earliest-born young of the year can be up to 4 months old and reproductively mature while their later-born part-siblings are still helpless nestlings. By the end of a highly productive breeding season the generations of weasels overlap, muddling the frequency distribution of age classes.

This extraordinary system is a huge advantage in the frozen north, where the small mustelids evolved (Sect. 4.5), because it enabled weasels to respond immediately to a sudden irruption of voles and lemmings. Females could rapidly produce a huge number of young ready to disperse far and wide, which explains the sudden short-lived irruption of weasels when they were first transported to New Zealand (Sect. 14.4). During a lemming irruption, weasels could also provide for coming hard times by collecting carcases in a cold-storage cache. Unfortunately, the same behaviour that is such an advantage in the Arctic becomes a liability in temperate countries where the instinct remains to kill many more rodents or chicks than needed, but the caches decay.

This flexibility makes the calculation of mean fertility and mortality rates especially difficult in weasels. The physical distinction between young and adult blurs quickly, and never can be defined in reproductive terms because the young mature so quickly. The compensation is, however, that these small animals are short-lived and their reproductive cycle is rapid. The young are usually vastly in the majority, and for most of the year can be distinguished from the smaller numbers of adults captured in the same month of their second or (unusual) third year. In many studies, division into two annual age classes, while not ideal, may be enough. Only the biggest kill-trapped samples can be large enough to have a high probability of including some of the rare older individuals.

Mortality Rates In theory, the use of kill-trapped samples to estimate the age structure of a population should make it impossible to distinguish natural from trap-induced mortality, and that could invalidate further analyses. Ideally, it would be better to use live-trapped samples, but live trapping studies always include fewer animals, and observations usually have to stop before the last of the marked young died of natural causes. Moreover, their disappearance can't be distinguished from emigration.

In practice, the effect of trapping history in removing the older weasels from a sample collected from gamekeepers is inconsequential, because they are so short-lived that the few older animals make rather little difference to the general age structure of the population and none to its dynamics. By contrast, the number of young produced each year is easy to assess from carcases, and much more significant, because the variation in productivity, juvenile survival and numbers is certainly controlled by food supplies rather than by trapping.

Can we use a list of ages of individuals to estimate how long weasels live? This is not an easy question to answer. The average or median age of a given wild population is simple to calculate, but that tells us nothing about the variation in distribution of the key age groups (juvenile, subadult, adult) across the total range. Even a list of the mean ages of several samples with standard deviations tells us little about the consequences of that age distribution for the dynamics of the populations concerned.

What a population biologist really wants to know is the probability of a defined group of animals surviving to particular ages, not just to the oldest age. For a given group, either the total population or a subset of it, longevity can be estimated in terms of survival probabilities, and their converse, mortality rates. The probabilities of survival and mortality also change under different conditions, as can be shown in a life table.

Life insurance companies use human life tables to estimate the risk of offering cover to people of different cohorts. Animal biologists took up the idea as a way of estimating the variation in survival rates (and, conversely, the mortality rates) among the age classes in a population. A frequency distribution of age classes for a stationary population (neither increasing nor decreasing over years) is proportional to the probability of each individual living to each age, *provided that* the population density is not changing, and that the age structure is stable. Neither condition is ever met by weasel populations.

In good years, the young weasels entering the population vastly outnumber the breeding adults, but scarcely match them, or even fail to appear altogether, in poor seasons. Further, all data on age distributions derived from trapping can sample only weasels that survive at least to trappable age, so they do not account for young that die at or before birth or in the den. Yet pre-independence mortality is probably the most important factor controlling the age structure, because in years of poor food supplies, mother weasels hunting alone struggle to find enough food to feed all, or even any of their young. If the total number of weasels born could be known, estimates of the mean ages at death would be much lower, and of the mortality rate of the first-year class much higher, than can be estimated from trapped animals.

Longevity Allowing for all those caveats, my sample of 455 British weasels eventually yielded a life table suggesting that the first-year mortality rate was 80% in males, 75% in females. This figure was surprisingly close to what would be expected from small animals that live their lives at such a hectic pace. Most of them had lived for less than a year, so the all the samples were dominated by the first year class. The average expectation of further life for a young weasel at independence was about 10 months in both sexes. The density of a population must therefore depend heavily on the production and survival of last season's crop of young.

In any samples collected from gamekeepers, there will be a well-defined peak in the risk of mortality every year in spring, when keepers kill a lot of weasels and stoats to protect their game birds. Robbie McDonald (Sect. 11.4) used this abundant material to construct a comprehensive model of the population dynamics of British weasels and stoats (McDonald & Harris, 2002). His results clearly showed the high mortality rate of both species when the keepers were most active in spring. On the other hand, spring is a stressful time for weasels anyway, especially if food is short. In Wytham's protected population, never subject to regular trapping, several resident weasels observed for months died or disappeared in spring, and the age distribution of weasels from Wytham was not different from those on the game estates.

2.6 Putting It All Together

So what can we know about the lives of Britain's smallest native carnivores? Comparing the only two available sets of field data, from the Carron Valley (Lockie, 1966) and from Wytham, one conclusion jumped straight out: weasel numbers and behaviour depend largely on habitat and food supply. Weasels can live in good numbers under thick, matted ungrazed grass on a Scottish hillside full of field voles, where they could thrive on their small, fiercely defended territories. By contrast, only a few weasels could live on open-floored English woodland inhabited by many fewer and smaller rodents. The Wytham weasels survived on the edge of viability even on their large, undefended home ranges. The average annual mortality rate of weasels protected from kill-trapping in Wytham was no lower than on a typical game estate where the local gamekeeper attempted to remove them from pheasant breeding areas.

Weasels (and by implication, all small mustelids) provided an early example of an animal population controlled from the bottom up, by food supplies and juvenile survival, rather than from the top down, by additive mortality due to traps and raptors. In due course, this conclusion was severely tested in the very different ecological landscape of New Zealand.

References

Crowcroft, P. (1991). *Elton's ecologists: A history of the Bureau of Animal Population*. University of Chicago Press.

Erlinge, S. (1974). Distribution, territoriality and numbers of the weasel *Mustela nivalis* in relation to prey abundance. *Oikos, 25*(3), 308–314.

Gorini, L., Wedul, S. J., Arnemo, J. M., Linnell, J. C. D., Boitani, L., & Nilson, E. B. (2013). Field anesthesia of least weasels (*Mustela nivalis nivalis*) with isoflurane. *Wildlife Biology in Practice, 9*, 7–13.

Hayward, G. F., & Phillipson, J. (1969). Community structure and functional role of small mammals in ecosystems. In D. M. Stoddart (Ed.), *Ecology of small mammals* (pp. 135–211). Chapman & Hall. https://doi.org/10.1007/978-94-009-5772-5_4

King, C. M. (1973). A system for trapping and handling live weasels in the field. *Journal of Zoology, 171*, 458–464.

King, C. M. (1975). The sex ratio of trapped weasels (*Mustela nivalis*). *Mammal Review, 5*(1), 1–8.

King, C. M. (1980a). Age determination in the weasel (*Mustela nivalis*) in relation to the development of the skull. *Zeitschrift fur Saugetierkunde, 45*(3), 153–173.

King, C. M. (1980b). Population biology of the weasel *Mustela nivalis* on British game estates. *Holarctic Ecology, 3*, 160–168.

King, C. M. (1980c). The weasel *Mustela nivalis* and its prey in an English woodland. *Journal of Animal Ecology, 49*, 127–159.

Linn, I., & Day, M. G. (1966). Identification of individual weasels *Mustela nivalis* using the ventral pelage pattern. *Journal of Zoology, 148*, 583–585.

Lockie, J. D. (1966). Territory in small carnivores. *Symposia of the Zoological Society of London, 18*, 143–165.

McCleery, R. H., Clobert, J., Julliard, R., & Perrins, C. M. (1996). Nest predation and delayed cost of reproduction in the great tit. *Journal of Animal Ecology, 65*, 96–104.

McDonald, R. A., & Harris, S. (2002). Population biology of stoats *Mustela erminea* and weasels *Mustela nivalis* on game estates in Great Britain. *Journal of Applied Ecology, 39*, 793–805.

Savill, P., Perrins, C., Kirby, K., & Fisher, N. (Eds.). (2011). *Wytham Woods: Oxford's ecological laboratory*. Oxford University Press.

Southern, H. N., & Lowe, V. P. W. (1968). The pattern of distribution of prey and predation in tawny owl territories. *Journal of Animal Ecology, 37*, 75–97.

Southern, H. N., & Lowe, V. P. W. (1982). Predation by tawny owls (*Strix aluco*) on bank voles (*Clethrionomys glareolus*) and wood mice (*Apodemus sylvaticus*). *Journal of Zoology (London), 198*, 83–102.

Part II

Part II First Understand Your Target

DSIR Ecology Division, 1971–77

3

Abstract

The Ecology Division of the former Department of Scientific and Industrial Research (DSIR) ran a forest ecology programme in the Orongorongo Valley. Among many co-operating projects was a single long line of 116 rodent trap run at quarterly intervals, providing baseline index data on the rodents available for stoats. The Division's technician designed the Edgar live trap for stoats and built 51 of them, and from late 1971 the same field methods that had worked well for Wytham weasels was applied to Orongorongo stoats. We recorded 29 captures of five stoats in the first 4 months, but only another 15 captures of four stoats over the next 3 years. The project was abandoned in early 1975. Meanwhile, the New Zealand Wildlife Service (NZWS) was becoming concerned about the declining numbers of an endangered endemic bird, the takahē, for which a strictly protected Special Area had been established in Fiordland. In May 1972, NZWS supported an expedition to establish a permanent array of humane Fenn traps to clear stoats from the bird's last refuge. The first two stoats killed were both fertilised females, and hundreds more have since been removed.

Keywords

Orongorongo Valley · Edgar trap · Live trapping stoats · Takahē Valley · Fenn trap

3.1 The Orongorongo Valley

The former Department of Scientific and Industrial Research (DSIR) was a Government sponsored research organisation comparable with the much larger Australian CSIRO. Its Ecology Division (Sect. 1.3.3) employed a team of specialist ecologists to run a forest ecology programme in the Orongorongo Valley, east of Wellington.

C. M. King, *Stoat in the Dock*, Fascinating Life Sciences,
https://doi.org/10.1007/978-981-96-4562-6_3

The Orongorongo Valley occupies a fold in the southern Rimutaka Range, to the east of Ecology Division's office in Lower Hutt. DSIR's Orongorongo Valley Field Station (OVFS) was originally built by the NZ Forest Service to study the ecology of Australian brushtail possums, an introduced fur-bearing marsupial. Ecology Division had taken it over and vastly expanded its facilities and staffing. My first brief on moving to New Zealand in 1971 was to join the team working there, and to contribute the same sort of information about the place of stoats in the forest ecosystem that I had previously studied in the weasels in Wytham (Chap. 2).

The Orongorongo Team Ecology Division's Field Station occupied a 1215 ha block stretching for about 6 km along the east bank of the river, covered with unlogged native lowland forest. It extended to the top of the nearest ridge (up to 840 m elevation) where the lowland tree species gave way to a forest of nearly pure southern silver beech (Nothofagaceae). The valley sides were steep, and the surrounding terrain was much broken by earthquake action and erosion. The only access was by Landrover along a rough track north from the coast, or along the shingle bed of the river. Otherwise, all transport throughout the area was by foot track.

The research team included specialists in the population ecology of most of the resident mammals, birds and insects, and of the forest vegetation. Irregular mast seeding of the forest trees produced sudden pulses of extra food for ship rats (*Rattus rattus*), feral house mice (*Mus musculus*), stoats (*Mustela erminea*), seed-eating birds and possums (*Trichosurus vulpecula*). Just as at Wytham, all members of the team routinely shared their data in order to advance our collective understanding of the animals, plants and their habitats as part of an integrated ecosystem (Fitzgerald & Gibb, 2001). One member of the team, Bob Brockie, has provided an eloquent and well illustrated account of the history of the forest and the achievements of the team based at OVFS up to 1992 (Brockie, 1992) .

The most important food resources for stoats everywhere are small mammals, which at OVFS included only wild house mice, ship rats, possum carrion and rabbits (*Oryctolagus cuniculus*). Rats and mice were hugely affected by beech masting (seedfall events) and the following surges of litter invertebrates, either in the mixed forest at river level or, especially, in the silver beech forest along the ridge tops. In poor seedfall years in the intensive study area around the field station, the average population density index of mice (so-called, although it is really only an index of abundance) was usually around 0–2% trap capture rate. After the huge beech seedfall in autumn (March–May) 1971, the mouse density index shot up to 6.1 in midwinter (August), and further to 11.2 in spring (November) (Choquenot & Ruscoe, 2000; Fitzgerald et al., 2004). Later studies confirmed that 1971 was a huge seedfall and mouse year throughout the country.

Moderate falls recur on average every 3–4 years. The size of the seedfall in autumn is now so firmly correlated with the abundance of litter invertebrates and rodents over the next 6 months that it can be used as a proxy for rodent abundance next spring (Alley et al., 2001).

Running through the lowland forest for the full length of the research block (6 km) was a single long line of 116 rodent trap stations at 50 m intervals, established by Mike Fitzgerald in August 1971 and run for 3 consecutive nights at quarterly intervals. Mike and his indefatigable technician Brian Karl (with occasional help from me) ran the long line for 25 years, and also monitored the feral cats that hunted the rodents and rabbits. A similar line of 36 traps monitored the rodents living on top of the main ridge, for a much shorter period.

Whenever Ecology Division's Director John Gibb could get away from the office, he observed the home ranges and behaviour of the rabbits that lived among the patches of vegetation on the gravel river bed, which shifted or were swept away with every spate and regrew when the flood waters drained. All members of the team contributed to a register of sightings of anything that could be of interest to the others. Chance sightings of stoats by observant colleagues doing their own work in the forest added many more data points to a preliminary estimate of the variation in numbers of stoats than could be got from my trapping records alone.

First Stoat Project It was already well known that the numbers and food habits of stoats would be strongly influenced by the chain of events following a heavy seedfall. Post seedfall irruptions are an easily monitored indicator of the huge pulses of resources added to the food supply of stoats through a predictable masting cycle, which collectively have disproportionate effects on stoat productivity and population dynamics. Most work on stoats in subsequent years was focussed on documenting these important events and their consequences. But at OVFS the initial task was to determine the numbers and food habits of stoats resident on the research block, relative to the data on changes in the populations of birds and forest rodents supplied by other members of the team.

The first requirement was to catch the resident stoats alive as often as possible, which meant establishing a network of humane and reliable live traps. The only traps available for stoats in New Zealand at the time were leg-hold (gin) traps, which did not kill it immediately but caused cruel injuries (Sect. 4.2). Fortunately, the Division employed a brilliant technician, Ray Edgar, and supplied his workshop with everything he needed to make (among many other things) wooden live traps for stoats.

The first trap line followed a circular course 1.4 km long around the field station, lured with rabbit gut and baited with white mice, and checked twice daily on foot for 1 week in every three. The traps were operated according to proven Wytham methods—the captives were anaesthetised , scats and fleas collected, teeth and reproductive condition examined, all according to techniques that had worked well before (Sect. 2.1). Captured stoats were marked with ear-tags, although there was no Plan B for identification because, unlike weasels, the line between a stoat's brown and white fur is always straight.

In the first 4 months of the study, from January to April 1972, a respectable total of 29 captures of five individual stoats appeared in 1159 trap rounds. Four of these stoats lived in the area around the field station, and one further away. One male first caught in January was recaptured 11 times before he disappeared in April. The second was a female, also first trapped in January, which was caught 8 times over the next couple of months, until she was found dead in a rat trap set up a tree (stoats can climb trees as well as any squirrel). The third was a male caught 7 times before disappearing in April, but then he turned up, still wearing his tag, about 500 m away in September 1974, when he must have been almost 3 years old. The fates of the others were unknown.

Setbacks Such a good start was immensely encouraging, but it did not last. By November–December 1972, stoats were so few that traps had to be more widely spread—too far apart for daily inspections on foot. Twenty traps were laid over 4 km along the vehicle access track, checked twice daily from the Landrover. In 668 trap-nights over that second summer, only two stoats were caught, one three times and one four times. Such dismal results were never going to give useable results, so they did not get beyond an unpublished internal report.

After a long pause for reconsidering, and experimenting with recording footprints (Sect. 9.1) and other alternative ideas, Ray and I together developed an improved version of a traditional drop-door trap with a large hole in one side fitted with a sliding door. When open, the hole led to a detachable nest box big enough to hold a substantial warm bed. The idea was that, if a captured stoat could be kept warm and fed for up to 24 h, it would be safe to inspect the live traps only once a day, hence permitting fewer checks and longer trap lines.

The addition of the nest box also had the advantage that, when the sliding door was closed with a captive inside, it could be removed and used as an anaesthetising chamber (Fig. 3.1a–c). We called it the Edgar trap, and Ray made precise technical drawings of every detail (King & Edgar, 1977). In August 1974 he made 51 of these, which were set for 5 months along a 5.5 km line along the track from the field station to the boundary of the research block. In 921 trap nights two stoats were caught, both in December, once each.

The key to further progress was to make trap inspections fast enough to permit a larger array. For example, if it were ever possible to tell from a distance which traps were still open, there would be no need to spend time turning aside to check them.

This was not at all easy in the days long before wireless technology. One idea was to attach a horizontal mercury switch to the trap door. When the door dropped, it would report the circuit interrupted and the trap occupied. The problem was that we had to lay lengths of wire to each trap from the nearest inspection point. It sounds clunky, and it was, but it was never implemented, because we couldn't get enough wire.

In total, 36 stoats were caught and marked at OVFS over 2748 trap rounds, plus 44 rats, five hedgehogs, two possums, one cat and one rabbit. Another 14 stoats were caught in rat or possum traps, and five were found dead. The first 29 of the 36 stoats appeared during the first 1159 rounds between January and April 1972, during what we later identified as part of a nation-wide post-seedfall irruption. That was a reasonable capture rate, but short-lived. It took another 1589 rounds to catch the other seven. Only years later could we work out why the post-seedfall cohort of

Fig. 3.1 (**a**) Extracting a live, anaesthetised stoat from the nesting-cum-anesthetising box. Orongorongo Valley, 1972. (Photo by Author). (**b**) Recording capture details on one of the few female stoats caught in the Orongorongo Valley, 1972. (Photo by Author). (**c**) Weighing a live, unconscious male stoat. (Photos by Author, all from Orongorongo Valley, 1972)

stoats disappeared so quickly (Chap. 11). Only seven of 51 food items identified were birds, far too few to answer any questions the team wanted to ask about the effects of stoat predation on forest birds.

Plan B Capture records of mice from the long line of rodent traps, and the number of casual sightings of stoats, both peaked in November 1971. Both declined to a low in May 1973. It became obvious that the good results in the summer of 1971–72 were due only to the short-lived pulse in extra food provided by the 1971 seedfall in the silver beech forest on the ridge above the station that had happened, by chance, only a few months before the first stoat traps were deployed. All unknowingly, those first traps had been set during a classic post-seedfall summer (Sect. 5.3). After that temporary irruption passed, the chances of collecting any statistically useful data on stoats at OVFS dropped to zero. The original aim of contributing information about them to the team's collective description of the forest ecosystem became impossible.

The problem was simply one of geography and scale: stoats are small, fast-moving and wide-ranging animals, which could not at that time (long before the advent of radio-collaring) be observed directly. To have any chance of catching enough of them alive, box traps have to be spread too widely to permit daily inspections on foot, which means that the trapper has to have some kind of assisted transport. But the valley sides were too steep, and the forest too thick, for any kind of vehicle to help. The project was abandoned in January 1975, and its final report was never published.

When faced with the same problem at Wytham, I had found that I could produce information from large numbers of carcases that could then be used to help understand a smaller number of live animals. John Gibb, always fair even when he disagreed, gave me permission to try. So began the National Parks Stoat Survey (Chap. 4).

3.2 Takahē Valley

Takahē Valley is a high-elevation hanging valley in Fiordland National Park, west of Lake Te Anau, the last refuge of the takahē, a rare endemic bird that had been believed until recently to be already extinct (Fig. 14.1). It had been rediscovered 1948, and strictly protected from human disturbance ever since, but its numbers were still declining. Early attempts to get rid of stoats had to use old-fashioned gin (leg-hold) traps that were still legal in New Zealand, and widely used to catch possums, rats and stoats. The precious birds continued to disappear.

The New Zealand Wildlife Service decided that the takahē needed better protection from stoats, so they recruited Jim Mills and Roger Lavers to set up a rescue operation. Early in 1972 I was asked to travel with them to Takahē Valley and advise on how to improve trapping operations against stoats.

Introducing Fenn Traps Gin traps cause fearful injuries to captured animals (Sect. 4.2), so I introduced the Fenn, a spring trap invented in Britain after gin traps were made illegal there in 1958. Fenns were designed to be placed in a tunnel barely wider than the movement of the jaws, which are then forced to close across the spine rather than round the leg. Fenns were the most humane traps available at the time. When correctly set, Fenns were supposed to kill stoats and rats quickly (Fig. 3.2).

The Fenn factory airlifted a box of Fenns around the globe in May 1972, just in time so that I could take them to Takahē Valley, and train the local parks and wildlife officers on how to use them. By the end of the 10 day trip, a trapping programme was established which caught two female stoats, and has since been widely extended (Hegg et al., 2012). Examining the two carcases quickly showed that the same methods of processing carcases that I had for previously developed for weasels (Sect. 2.3) also applied to stoats. My field class instructing the Wildlife Service rangers on the technique of how to age and sex the carcases (Sect. 4.6) was an added bonus.

The Wildlife Service's boss lamented that the two female carcases were probably the most expensive samples ever collected. That might have been true, but they were

Fig. 3.2 The first Fenn trap set in May 1972 killed the first female stoat removed from Takahē Valley. (Photo by Jim Mills)

also the most significant, because they were certainly already fertilised (Sect. 4.5), so had special value to the ongoing programme to protect the takahē. The trapping programme has continued ever since. For further details see Sect. 14.2.1.

References

Alley, J. C., Berben, P. H., Dugdale, J. S., Fitzgerald, B. M., Knightbridge, P. I., Meads, M. J., & Webster, R. A. (2001). Responses of litter-dwelling arthropods and house mice to beech seeding in the Orongorongo Valley, New Zealand. *Journal of the Royal Society of New Zealand, 31*, 425–452.

Brockie, R. (1992). *A living New Zealand forest*. Bateman.

Choquenot, D., & Ruscoe, W. A. (2000). Mouse population eruptions in New Zealand forests; the role of population density and seedfall. *Journal of Animal Ecology, 69*, 1058–1070.

Fitzgerald, B. M., & Gibb, J. A. (2001). Introduced mammals in a New Zealand forest: Long-term research in the Orongorongo Valley. *Biological Conservation, 99*, 97–108.

Fitzgerald, B. M., Efford, M. G., & Karl, B. J. (2004). Breeding of house mice and the mast seeding of southern beeches in the Orongorongo Valley, New Zealand. *New Zealand Journal of Zoology, 31*, 167–184.

Hegg, D., Greaves, G. J., Maxwell, J. M., Mackenzie, D. I., & Jamieson, I. G. (2012). Demography of takahe (*Porphyrio hochstetteri*) in Fiordland: Environmental factors and management affect survival and breeding success. *New Zealand Journal of Ecology, 36*, 75–89.

King, C. M., & Edgar, R. L. (1977). Techniques for trapping and tracking stoats (*Mustela erminea*); a review, and a new system. *New Zealand Journal of Zoology, 4*, 193–212.

The National Parks Stoat Survey, 1972–76

4

Abstract

Given a large enough total sample and a wide range of replicates, it is possible to learn a lot about basic stoat biology from simple routine dissections. This survey collected 1599 stoats, plus 40 weasels and 56 ferrets from 14 study areas representing native southern beech (Nothofagaceae) forests, podocarp/broadleaved forests, and tussock grasslands, at both low and high elevations and across latitudes from 38°S to 45°S. We found that stoats can survive injury by inhumane traps or vehicle impacts; they eat whatever they can catch, and survive on large native invertebrates when necessary; their black-tipped tails help them avoid swooping raptor talons; males grow larger than females, especially when well fed as nestlings, and larger males can better fight off rivals while smaller females can be more efficient hunters; stoats in New Zealand have become slightly smaller than their ancestors; fecundity of females (ovulation rate) is fixed a year in advance of littering but fertility (litter size) is finely adjusted to maternal condition by intra-uterine mortality over winter; skull shape and bacula weight reliably change with age; seasonal changes in sex ratio of trap captures is related to the annual cycles of behaviour of individuals; traps set too wide miss many females.

Keywords

Stoat diets · Body size/prey size link · Delayed implantation · Sexual dimorphism · Sex ratio of captured stoats

4.1 Organising the Collections

By November 1972, 1 year on from finding that live-trapping stoats in the Orongorongo Valley was proving untenable (Chap. 3), I turned to an alternative project, which had worked before (Sect. 2.3). The National Parks Act of 1952

C. M. King, *Stoat in the Dock*, Fascinating Life Sciences,
https://doi.org/10.1007/978-981-96-4562-6_4

Fig. 4.1 A National Parks ranger inspecting a Fenn trap set in a tunnel on the way to Milford Sound, Eglinton Valley, 1972. (Photo by Author)

required park boards to preserve the native fauna and flora, and to exterminate introduced species in their parks. So carcases would be easy to collect from park staff who were already routinely killing and discarding stoats as part of their normal duties (Fig. 4.1).

This was not a new idea, since Mike Fitzgerald had done a similar study in 1964 (Sect. 1.3.5), but the development of new methods of analysis made a repeat seem worthwhile. A good opportunity to collect material arose in October 1972 when the Chief Rangers Forum, attended by all the chief rangers of all the then 10 New Zealand national parks, was held at Totaranui, in the Abel Tasman National Park.

All the Chief Rangers agreed to help, and the National Parks Authority generously supported Ecology Division to make a preliminary survey of stoat biology at national scale. Another four areas were added as opportunities arose, of which the best was Craigieburn Forest Park—a high elevation area with no rats. New Zealand Forest Service staff offered to check traps daily as a means of reducing the boredom of driving around a fixed route inspecting meteorological stations.

That was the start of what turned out to be a commitment of 10 years of field, laboratory and desk work on stoats collected from around the country. Many different sources offered help for much of that time. It was also huge advantage be able to apply the same methods previously evolved for carcase analyses of weasels in Wytham (Chap. 2), or build new ones upon them, so once the dead stoats started arriving from the national parks, the lab procedures could start like clockwork.

Lab Procedures Parks staff agreed to replace the usual leg-hold traps with imported Fenns, which, besides being supposedly more humane, were better for research than gin traps. The killing action of the Fenn ensured that most captured stoats had died quickly (but see below, Purdey et al. (2004)), so the rangers could provide clean specimens complete with gut contents and undamaged skulls and teeth (Sect. 3.2) as part of their normal field routines (Fig. 4.1). Supplies of a printed label to be attached to each carcase ensured that the vital details of capture date and location were recorded. In exchange, whatever information could be deduced about how stoat populations work in their particular areas was returned to them. The rangers sent in so many boxes of stoats that this apparently simple request turned into a huge project.

Ray Edgar (Chap. 3) made a set of insulated boxes, one for each area, which were air-lifted to the parks and back to the lab in Lower Hutt. The 14 sample areas represented a wide range of habitats, from sea level to 1400 m elevation, from 1100 to 5100 mm annual rainfall, from 38 to 45 degrees of latitude, and including samples of most major classes of natural vegetation (King & Moody, 1982).

By mid 1974, the numbers of animals arriving for processing (eventually 1599 stoats, plus 40 weasels and 56 ferrets) had become a two-person job. The National Parks Authority employed technician Jenny Moody to help me. Between us we put the carcases through a routine process which included weights and measurements, records of habitat and capture method, moult status (classified from the inside of the skin), reproductive status, gut contents, parasites, and clean skulls. In addition, it took 3 months in the histology lab at Victoria University to make serial sections of 927 ovaries (including 439 pairs), from which we were eventually able to calculate the annual variations in ovulation rate of female stoats (Fig. 4.5) (King et al., 2003).

Because of the way the survey was set up, we had no control over how and when the animals were collected. That made the collections quite heterogenous, and the normal field protocol requirements specified by statisticians impossible to meet. There was no fixed sampling regime, no replicates and no before/after comparisons, so purists would (and some did) dismiss the results as meaningless stamp-collecting. Of course, that view was quite understandable, and I sympathised with it.

But there was no other way to collect such a very large total sample, representing enough local subsamples to allow preliminary comparisons between groups of stoats from different environments. And in practice, the uncontrolled variation in the material was seldom extensive enough to completely invalidate our analyses. The results therefore offered a simple description of the variability within and between samples in multiple parameters, especially in the hitherto unknown but essential details of the stoat reproductive cycle (Sect. 4.5). It was nothing like a proper demographic analysis, but, with all its faults it documented a lot of fundamental data that became important in supporting later population modelling (Sect. 11.4).

4.2 Inhumane Traps and Traffic Injuries

Because nearly all parks rangers and wildlife officers used Fenn traps (Fig. 3.2), but also collected stoats accidentally caught in leg-hold traps set for possums, it was possible to test whether the Fenn trap really was as humane as Mr. Fenn claimed.

Evidence of Injuries Of the 1599 stoats collected during the survey, 336 were labelled as having been caught in gin traps, of which 41% had gross external injuries (broken legs or teeth, crushed or severed teeth) and 32% were still alive when collected. By contrast, of the 966 caught in Fenn traps, 8% had external injuries and 4% were still alive. Many more of the gin-trapped animals had empty guts (28–13%), showing that they had lived long enough to defaecate; and more of the gin-trapped bodies still had fleas (18–11%) showing that they had remained warm for longer (King, 1981).

If part or all of a foot was found in the gut of a stoat whose body had the same part missing, it was considered to have been chewed off by the captive in attempting to escape. Of 29 stoats that had bitten off their own foot, 21 had been caught in gin traps. Six others had well-healed injuries showing that they had escaped alive from a previous trap, in which three had lost one whole front leg, two lost part of their tail and one lost four toes from a front foot. Because the front feet are the first to strike the trigger plate of a trap set in a tunnel, these distressing injuries were an understandable consequence of the way stoats tend to hunt by venturing headfirst into burrows.

The record of injured animals collected from the supposedly humane Fenn traps raised suspicions, confirmed in detail by one of few studies in which Fenn traps were inspected daily (Purdey et al., 2004). At least six of the 63 captured stoats were not immediately incapacitated by the Fenn traps even when caught correctly, by both the neck and lower abdomen. We found four still alive and attempting to escape from the tunnel. M1 had died only after dragging itself with the trap to the tunnel exit, and M4 after attempting to dig its way out. Others had chewed the tunnel interior extensively. These observations on wild stoats are few but unusual because most Fenn trap lines run in recent years have not been checked daily. They confirm laboratory tests showing that the Fenn trap (both M4 and M6 versions) is not humane enough to meet New Zealand's new ethical standards for kill traps.

But eventually Animal Welfare legislation set guidelines requiring all 10 test animals to be rendered irreversibly unconscious within 3 min. The Fenn could not pass these much tougher animal welfare tests (Warburton et al., 2008), so was outlawed. The Department of Conservation (DOC) then funded the development of much more advanced humane traps, the DOC 200 and 250 series that later became mandatory in New Zealand (Fig. 4.2). It was more humane but less useful for research, because its action was likely to crush the skull.

Fig. 4.2 A dead stoat in a DOC 250 trap. (Photo by Dave Campbell)

Vehicle Impacts A different picture emerged when we looked at the hind legs (King, 2021). Among 561 carcases from which both femurs had been extracted and cleaned (for another study, Sect. 4.6), we found five cases of grossly deformed healed fractures of femurs (Fig. 4.3). Another case was contributed by Andrew Veale from the monitoring programme on Rangitoto, an island 3 km offshore from Auckland (Sect. 12.1). How could these animals, caught by one or both hind legs, have wriggled out of either kind of trap and lived long enough to heal themselves?

For them, we eventually concluded that the sources of injury were not traps but vehicle impacts. Stoats do not despise carrion when other prey are scarce, and in the process of feeding on a roadkill, a stoat would be itself exposed to danger. The hind legs are the last part of the body to get out of the way of an oncoming vehicle, and it would not be surprising if some stoats left their escape a fraction of a second too late. The one showed in Fig. 4.3 was an elderly male that had lived next to the busy road to Milford Sound, a prolific source of carrion. I once found a dead possum on a road, with a dead stoat beside it.

Whatever the source of the injury, the question remains: How could an active solo hunter with (presumably) no social support, suffer a broken leg and still survive long enough to heal itself? The genetic identity of the Rangitoto stoat mentioned above showed that it must have come from the mainland, and was presumably injured there. Yet, even with a broken leg, it was still able to swim to the (traffic-free) island (swimming stoats paddle mostly with their front legs (Sect. 7.2), but still....). A stoat must surely be one of the toughest little animals alive.

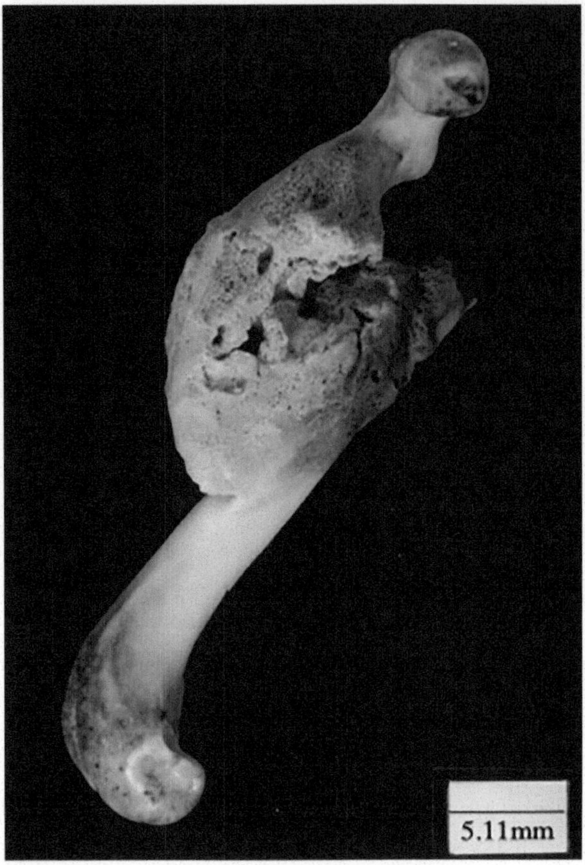

Fig. 4.3 Lateral perspective of the left femur of FL132, a male from the Eglinton Valley, born October 1968, killed 17 July 1973, aged 57 months. The two ends of the femur, presumably broken by a vehicle impact, had been knitted together out of line during life, long before the same animal was killed in a trap. (Photo from King, 2021)

4.3 Stoat Diets

A Different Menu The stoats and weasels transported from Britain to New Zealand in the nineteenth century joined a simple but un-natural community of introduced mammals (King et al., 2017). The range of body sizes of prey available to the new arrivals was different from anywhere in the northern hemisphere: fewer small mammals of mouse size (up to 30 g), but relatively more of rat (80–200 g) and rabbit size (500 g to 2 kg), plus abundant large native invertebrates. Australian brushtail possums, introduced in the late nineteenth century for their fur, average some 2–3 kg, and provided an important source of carrion along roads and fur-trapper lines. The only competing predators were feral cats, ferrets, and the native morepork owls, harriers and bush falcons.

Weasels, however, were (and still are) specialist hunters of small rodents (Chap. 2), so are less adept at hunting grown rabbits than are stoats. They soon cleaned out all available prey of small size (wild house mice, native lizards, and large native invertebrates) when they first arrived, and then could not find enough substitutes for their normal prey, the voles, native mice and lemmings of the northern hemisphere. New Zealand has proved inhospitable for weasels: They are not extinct, but they are scarce and patchy in their distribution compared with stoats (Vattiato et al., 2023).

Interpreting Results The food remains of stoats in national parks were easy to identify because (as the Orongorongo study had already shown) the range of potential prey animals in New Zealand is small, and the hairs of the few mammals eaten could be distinguished without the sectioning process that was necessary in UK. But interpreting the results is another matter. Presenting diet data in percentages requires only a simple calculation, but conceals a hidden trap for young players.

The important question to decide is, is this item a percentage of (1) the number of items identified, or (2) the number of gut or scat samples examined? Option (1) means that any given item can be assessed only in relation to all other items, so adding or subtracting one would change all the other results without reference to what the stoat itself is doing. We avoided this method when presenting general comparisons between study areas, years and seasons.

Option (2) requires each category to be recorded as present or absent in each gut (or scat) examined, including the 'empty' ones scored as zero, and the number of positive scores calculated as a percentage of the total number of guts or scats containing food. Only thus can the distributions of each prey category be compared between samples independently, without reference to the rest of the diet.

On the other hand, stoats are part of a dynamic animal community whose components interact with each other, and changes in numbers of any one prey species often do ripple through the network. Hence the dietary categories visible to us were not necessarily as independent in the biological sense as in the statistical sense.

The question of greatest interest to the rangers supplying the dead stoats, which they worked hard to remove in order to protect native fauna, was: how many birds are killed by stoats? what was the effect of these alien predators on bird life? That was to them a simple question, and they were disappointed to discover that there is no simple answer. At this stage in the project, all that could be done was to tabulate the contents of the 1514 guts of stoats we examined. There is no way to make the leap from diet to predatory impact. Even at local scale, with prey data available, extrapolations from general averages can be misleading, because individual stoats have personalities that determine their hunting strategies, and even develop their own specialities (McAulay et al., 2021). Young stoats just out of the nest hunt in a pack with their mother for several weeks, and learn from her what potential prey are good to eat. Individual preferences set at that early age apparently remain for life (Johnstone & Garvey, 2023).

Only 1250 of the 1514 guts contained any identifiable food, totalling 2643 prey items (King & Moody, 1982: Part II: 57–80). They were divided into nine major categories, as follows: mice were found in 19% of 1250 guts, rats in 6%, lagomorphs (technically, this category includes both rabbits and hares, but in this context very few if any would have been hares), 18%, possums 10%, birds 43%, lizards (geckoes and skinks) 5%, invertebrates a whopping 41%, freshwater crayfish 2%, and carrion about 1–2% (probably an underestimate, since carrion is difficult to identify). Remarkably, this broad outline of results was very similar to that which had been provided by Mike Fitzgerald in 1964 (Sect. 1.3).

Every honest list of such categories also has to have one labelled "Unidentified", which in this case usually referred to possums or rats. This was predictable, because (1) the distribution of unidentified hairs paralleled that of possums in the general results, and (2) few could have been lagomorphs because even a few long hairs from a rabbit were easily distinguishable by their unique medulla pattern. Mouthfuls of fur can be avoided by a stoat stripping meat from a possum or rat carcase, and the fur of mice was recognisable from texture and length, so these items were easy to spot.

There must also be a separate category for empty guts (containing nothing, or only bait or stoat fur). These items amounted to 17% of the 1514 guts examined, and they were excluded from further consideration. There also had to be a way of excluding possums and rabbits that had been provided by the trapper as bait.

Obviously, such a gross lumping of results gave only a broad view, disguising a lot of local, seasonal and individual variation, so the next step was to examine more closely the 866 animals collected in forests, and the 146 from Mount Cook, the only large sample from tussock grassland. Both offered the same unsurprising conclusion: stoats eat whatever they can catch.

Effects of Age and Sex The forests of late–nineteenth-century New Zealand were well stocked with food for stoats, provided they could adapt their hunting techniques to the unfamiliar range of prey sizes available. Forests also offered them shelter from the harrier hawks, cats and ferrets that preferred more open country. Not surprisingly, the earliest stoats to arrive moved into the forests at once. Even though their new environment contained a radically different array of potential meals compared with UK, stoats adjusted quickly and thrived.

By the time young stoats are old enough to be trapped, they can kill as well as can the adults. So there was no difference in diet with age, and stoats of all ages were considered together. But a difference between the diets of males and females is more likely, because male stoats are at least twice as large by weight than females, and there is a general correlation between the body sizes of a predator and its usual prey. That meant that the prey results of the two sexes had to be inspected separately.

In all seasons, female stoats ate more small items such as mice and invertebrates. Males ate more large items, especially possums and 'unidentified', which go together because possums were easily available as carrion discarded by fur-trappers, or as road kills (Sect. 4.2).

On the other hand, hunting stoats are bold out of all proportion to their size, so the females could still kill rats and rabbits when mice were scarce. In fact, both sexes of stoats ate rabbits, birds, rats and insects equally often. So the reason that males took more possums was probably because they had larger appetites, not because females could not tackle larger prey.

Stoats are efficient hunters, and the details of their menu depends largely on the habitat where they most often hunt, the season of the collection, and what their mothers taught them counted as food. So the general survey found differences in diets between stoats collected in different forest types and seasons, which probably reflected most of the real changes in the different prey species that lived there. For example, possums are more common in lowland podocarp forests than in alpine beech forests, whereas mice have the opposite preferences, and rabbits generally avoid forests of any sort, except along roads or forest edges.

Birds could not be identified to species, so they all had to grouped into a single category. Cleaning out nests is an easy way to for a lucky stoat get a good meal, so it was not surprising to find at least two sets of unidentified infant mammals. Stoats also take reptiles, as in one group of six skinks, and invertebrates, as in one collection of 14 tree wētā. One stoat ate 53 wasps in one sitting.

Preferences Stoats have a built-in list of dietary preferences, and mammals of whatever species are available locally are always at the top. In their native habitats stoats very rarely eat invertebrates, which is why the large proportion of such prey eaten by New Zealand stoats appears rather surprising. But undisturbed invertebrates in New Zealand can be large and nutritious, especially wētā, a native Orthopteran insect, which can grow large enough to be called 'an invertebrate mouse'. So these are not eaten by accident—rather, they are important starvation rations (Sect. 4.4.2).

When rodents are scarce, invertebrates can be the difference between survival and death. On two large rat-free Fiordland islands, invertebrates are life-savers for stoats, comprising 92% and 97% of the island stoats' diet in winter (Murphy et al., 2016). In general, stoats eat whatever they can catch, but young stoats develop their food preferences when first learning to hunt with their mother, which probably accounts for some individual variation in diets among adults. Some may never be interested in rodents, because they got early lessons on how to hunt crabs or wētā, but rodents would usually be a common item.

Conversely, when food is abundant, stoats will kill much more than they can eat immediately, and cache the rest. In cold boreal climates, this habit is a real advantage, because cached food can last well into the post-lemming-peak hungry years. Weasels do the same (Sect. 2.5). Unfortunately, the same habit also makes stoats particularly dangerous pests in New Zealand, where a stoat will kill every chick in a nest and remove them all before stopping to eat one (Little et al., 2017). In the high alpine refuges of the endemic rock wren, DOC's motion-activated cameras recorded several occasions when a stoat entered a wren's nest, killed all the chicks, and carried them away one by one (Fig. 14.3). This habit makes it possible for a single stoat to do a lot of damage, like the one that killed 90 adult diving petrels on Motuotau

Fig. 4.4 In open country, stoats kill many rabbits. (Photo by Patrick Garvey)

Island off Mt. Maunganui before it was trapped in 1996 (https://www.nzbirdsonline.org.nz), or the one that invaded the Orokonui Sanctuary (Sect. 13.2).

The Tasman Valley in Mount Cook National Park provided our only study area sampling a broad area of open grassland, and the stoats that lived there depended much more on rabbits than on mice or rats (Fig. 4.4). The ups and downs of rabbit numbers drove the Mount Cook stoat population through a timetable different from that of forest populations. For example, the summer of 1975/76 was a relatively good one for stoats at Mount Cook (3% trap success, the highest in 4 years) but a poor one in the Eglinton forest (0.7%). So not surprisingly, among the 146 Mount Cook guts we examined, 56% had eaten lagomorphs. This was an estimate rather than a firm number because there had to be a correction for the number of traps baited with rabbit meat. But that uncertainty could not disguise the fact that rabbits were the staple resource of Mount Cook stoats, just as they are in Britain. The next most frequent items of their diet were invertebrates (40%), birds (34%), and lizards (geckos 19% plus skinks 8%). Rats (1%) mice (6%) and possums (0), all important prey for stoats in forest habitats, were effectively absent at Mount Cook. A very similar list of items appeared in 206 scats collected at stoat dens in the same area decades later (Sect. 14.2.9). Again, the principal items were lagomorphs (56.8%), invertebrates (49%), and birds (36.4%) plus lizards (9.2%, nearly all skinks) and mice (6.8%).

Energy Value of Prey Because the main items of prey eaten by stoats varied widely in body size, not all items were of equal nutritional value. So another approach is to weight the main categories by body size, thereby producing a very different picture. This answer has statistical drawbacks, so can be useful only to identify which prey species were of greatest energy value. The answer was clear. When the numbers of prey items identified were converted into biomass, the larger mammals provided the most by far of the stoats' food requirements. A stoat could get more energy from 50 g of rat or rabbit meat free of indigestible fur and bones than from 50 g of mice. Invertebrates were eaten frequently but normally accounted for only about 10% of diet by volume. Without any information on prey resources, this early survey could obviously not be accounted a study of predation, but it provided useful background information for later studies that could.

The trappers were usually disappointed to hear that our general survey of stoat guts could not tell them what they most wanted to know: what was the impact of stoat predation on the numbers of birds in the forest? Alas, only the birds themselves could answer this question, and suitable opportunities to ask them did not arise until decades later (Chap. 14). Our data could say only that birds were the single most frequently identified category of prey, found at rates ranging from 15% to 62% of guts per quarter in females and 25–53% in males, depending on the season.

At that time, these grossly generalised data could offer no sign that that New Zealand stoats ate fewer birds at times when there was plenty of more preferred mammalian prey (unlike the Wytham weasels, Sect. 2.2). Neither did a preliminary comparison suggest that New Zealand stoats make up for the lack of voles (an important food for stoats in their native habitats) by eating more birds compared with those in UK. We had no option but to bottle the frustration and await more informative studies focussed on the birds (Chap. 14).

4.4 Body Size

Animals are not the size they are just by chance. Body size is one of the most important characters that determines an individual's life history, including what it can eat and who eats it; whether and where it can run, climb, or swim; where it can hide itself and its young; how soon it can breed, and how fast its young grow. The long thin bodies, short legs, rapid metabolism and fierce temperament of stoats and weasels are important in explaining how they survive as successful small predators in a world full of larger ones likely to see them as prey (King, 1989).

The Stoats' Black Tipped Tail Indeed, one of the most constant characteristics of the stoat, which always distinguishes it from the smaller but otherwise similar weasel, is the conspicuous black tip on the tail. Stoats are small enough to be confused with other prey by a stooping raptor or owl. So when stoats cannot avoid crossing an open space, they run like small bolts of brown lightning, flourishing their black-tipped tails behind them. Experiments with tame hawks and dummy model stoats with and without tails showed that an attack from the air could be deflected if the hawk focussed on and grasped the thin end of the tail rather than the body (Powell, 1982).

(Incidentally, this seem to be a strategy that works so well that it has evolved in another quite different small carnivore exposed to the same risks. The slender mongoose *Galerella sanguinea* belongs to an entirely unrelated South African family of small carnivores, the Herpestidae, but also lives among many aerial predators. It also has a long black-tipped tail).

The breed-fast-and-die-young lifestyle of all small mustelids explains why their numbers depend on the breeding success of the most recent season, so their populations swing wildly with food supplies from one season to the next. In New Zealand, a very different habitat from their northern hemisphere homelands, all these characters could be expected to determine their lives in unforeseen ways, with consequences critical for their numbers and predatory powers. So pinning down the details of the growth and body size of the stoats now living in New Zealand was always going to be an important question for the National Parks Stoat Survey.

Growth Rates The first stages of the lab processing of each stoat as it came in (Sect. 4.1) produced a lot of basic data, which we used to plot the growth curves of male and female stoats by DSB age class (defined in Sect. 2.4) and source population. The aim was to estimate their body size at maturity; compare the body size of adults collected from different habitats and years in relation to the different food resources available to them; and to compare the body sizes of stoats collected in the 1970s with those of their ancestors imported from Britain in the 1880s.

By the time the young first leave the nest at the age of 6–8 weeks, the males were already larger than their sisters. All the young grew rapidly until late summer, by which time the young females were essentially full grown (at 97% of the adult weight). Young males of the same age had got only as far as 80% of adult weight, and remained smaller than adults over winter. At the end of winter they suddenly put on a rapid adolescent growth spurt, and reached adult size by their first birthday in late spring (King & Moody, 1982: Part III: 81–102).

Most samples were large enough to reveal a surprising and significant degree of variation in stoat body size around New Zealand. The condylobasal length is a simple measure of the size of the skull, which, when plotted against body length, is a stable measure of body size. Both skull and body length are set by nutritional condition during growth, and they are thereafter not much influenced by daily variation or food supply. By contrast, body weight is variable with time since the last meal, as well as with nutritional condition which can be added or lost at any age.

The range of variation in local mean skull lengths for adult male stoats in New Zealand was huge (48.9–51.5 mm), almost as great as across western Europe (46.8–49.6 mm). The skull size of stoats collected from southern beech forests averaged larger than those from podocarp forests by 3–4%. The difference was significant in both sexes, and in both adults and young, and the same patterns reappeared, less crisply, if tail length or body weight were plotted in place of skull length. Since stoats had reached the South Island from Britain only in the 1880s and had been distributed without reference to body size, this remarkable geographic pattern seemed at first sight to have developed in less than 100 years.

The survey material included a disproportionate number of stoats collected in beech forests during post seedfall years, when mice, and stoats, were abundant. Closer analyses of the same material later found that young stoats born during post-seedfall peaks benefitted from enhanced growth associated with the additional food supplies available during their birth and nestling stages (Powell & King, 1997). So the average size of stoats in beech forest habitat in all years was boosted by the much larger samples we had collected during mouse peaks. Fortunately, the range of size variation across all the habitats we sampled was much larger than could be explained by this sampling error, so the conclusion, that the long-term mean stoat body size across New Zealand in all years is strongly influenced by habitat, remains intact.

4.4.1 Sexual Dimorphism

Sexual dimorphism, a universal trait in mustelids, is well established at an early age (King & Powell, 2007: table 4.1). In the general New Zealand population we sampled, adult male stoats were on average 9.6% larger than females, ranging locally from 6.5% to 12.0%. Why the difference? One explanation cites the sexual advantage of larger size to males in competition for females, combined with the opposite energy-saving advantage of small size to females for burrow-hunting rodents to feed her young. Another theory points to a reduction in competition between the sexes for food, because the larger males are better able to tackle larger prey capable of some self-defence, like full grown rabbits (Fig. 4.4). But any such advantage promoting separate diets could be the simple consequence of sexual dimorphism, not the explanation for it.

Roger Powell from North Carolina State University had a different hypothesis. He predicted that young male stoats born when mice are abundant should grow larger than males born in mouse-poor years, because good food supplies during the critical early growth phase can enhance the adult size of young males. Young male stoats are already larger than young females by the age of 2–3 months, which makes them more expensive to rear, and more sensitive to nutritional conditions during their juvenile dependency.

The same prediction should not apply to young females, because the theory expects their adult size to be constrained by the need to remain small enough to hunt rodents efficiently through their burrows and nests, while the reduced energy needs of their own small size enables them to pass more of their catch to their young. Hence, Roger concluded, the degree of sexual dimorphism in any given year should be linked to the extent of food supplies during the growth phase of the young. Nestling stoats are especially vulnerable to periods of food shortage, because until their fur grows, the young are dependent on her warmth. When the mother has to leave them alone while she goes hunting, they go into temporary cold rigour and stop growing (Segal, 1975). Conversely, young stoats grow faster and better when born into seasons of plenty (Powell & King, 1997).

I selected the jaws of adult stoats and sent them to Helen Grue, a tooth sectioning specialist in Denmark, and in due course the results came back (Sect. 5.1). Roger took all my records of stoats and mice through two beechmast cycles, and calculated age-specific linear regressions of body and skull length against mouse abundance.

All the skull length correlations were positive and significant, including those in the first age class. The problem was that although, as predicted, males definitely grew bigger when mice were abundant, so did the females. Therefore, sexual dimorphism did not differ in any size measure between stoats born in good versus poor years. Males more than 3 years old were similar in skull size regardless of birth year (Powell & King, 1997).

It is true that well-fed young males can grow large enough to have a temporary selective advantage in the current breeding season, because they can, presumably, outcompete small males for access to females. But there is no longer term effect on sexual dimorphism, because temporary variations in body and skull size are not a heritable character, only an index of current food supply for the young (Powell & King, 1997).

The smaller male stoats born in non-seed years live longer than larger males, because the mating advantage of the larger seed-year males is followed by a serious disadvantage in survival probability as mouse populations crash. In the 2–3 year intervals between seed years, the balance of selective advantages slips back against males with larger food requirements. If they live in an environment with relatively few large and meaty mammalian prey (rabbits and rats), they do less well on invertebrates than females can.

When the only small rodents available (wild house mice) are abundant, female stoats do indeed concentrate more on them than do the males, but in most years, house mice are not nearly large enough or numerous enough to live on (White & King, 2006). So female stoats must often tackle larger prey (rabbits and ship rats) as often as do males (Sect. 4.3.1). But the absence of the northern-hemisphere voles in New Zealand has removed the ancestral advantage of small size to burrow-hunting females in favour of the need to tackle larger prey when mice disappear.

It is possible that northern-hemisphere stoats and weasels born during vole or lemming peaks also respond to extra food supplies by growing larger, but no-one has collected large enough samples through a complete boom and bust cycle to find out.

In the end, a full explanation of sexual dimorphism has to take into account other characters, including that the female stoat lives and hunts alone, and has to support and defend her litters with no help from the male. The discussion continues (Noonan et al., 2016).

4.4.2 Body Size and Prey Size

One measure of the distribution of body sizes in a predator's diet is a 'prey size index'. A seminal paper published in 1987 proposed a strong correlation between the body sizes of stoats and of their main vertebrate prey (Erlinge, 1987). The smallest European stoats live in the far north, where being the right size to hunt small rodents under snow was a condition for survival and reproductive success in glacial times, and over-winter in the Arctic (Sect. 4.5). In Britain, home of the ancestors of New Zealand stoats, under-snow hunting is less of an advantage than is being large in competition for mates, so British stoats are the largest in Europe, and are specialist rabbit-hunters (McDonald et al., 2000).

Comparison with UK Ancestors New Zealand is a very different place from the native habitat of stoats . The shortage of small rodents and relative abundance of rats and rabbits skews the mean size index for *vertebrate prey* upwards. Therefore, in theory, the average size of stoats (despite huge local variation), should be slightly larger than that of their British ancestors. But in fact, stoats in New Zealand make up for the lack of small rodents by eating a lot of *invertebrates* (Sect. 4.3). The total mean prey size indices including invertebrates are *smaller* than for vertebrates alone, but still correlated with local variation in stoat body size (King, 1991a). Overall, selection for small size should have favoured smaller stoats in New Zealand over the 100 plus years since stoats were introduced.

The first attempt to detect any changes in the mean body size of stoats in New Zealand compared with their European ancestors used the only museum records of British stoat body weights available at the time, that were mostly rather old. The males averaged 321 g and females 213 g. The New Zealand males ranged from 284 to 356 g (mean 324 g), and females 195–226 g (mean 207 g), so at first glance it looked as if stoats had become slightly larger than their ancestors. But a later collection of British stoats made by Robbie McDonald in 1995–97 (McDonald & Harris, 2002) recorded higher averages of 367 g for males and 242 g for females in Britain.

The difference in the British data seems to be correlated with the date of specimen collection. The myxomatosis pandemic removed nearly all rabbits from the British countryside by the 1953. For years after that, stoats had to rely on small rodents (23% of diet) almost as much as on the few rabbits remaining (25%), and so stoat carcase weights remained atypically low (McDonald et al., 2000). By the 1990s, rabbits were abundant again (65% of stoat diet), small rodents were down to 16%, and stoats were recovering their previous larger average size, more likely to represent the 19th-century stoats sampled for transport to New Zealand.

The corrected comparison with British data published years later changes the picture. It shows that, except in beech forests during a post-seedfall summer, New Zealand stoats actually average slightly *smaller* than their ancestors (King et al., 2017). Updating early conclusions in the light of new data is a common and necessary experience in science.

4.5 Reproductive Cycles

Stoats not only display a huge difference between males and females in mating strategy and parental behaviour (sexual dimorphism), they also mate in spring but do not produce the consequent young for another 9–10 months (delayed implantation). Even more unusually, while young males mature at a year old, young females reach sexual maturity before they leave the nest (sexual bimaturism). These characters are absolutely constant, in all years, all habitats (King & Moody, 1982: Part IV: 103–118) and we need to understand them, because they have a significant effect on our options for the management of stoat populations.

Evolutionary Origins The evolution of this remarkable system began in the boreal regions of the northern hemisphere, the native habitat of the first stoats and their ancestors. They separated from their nearest relatives some 4.5 million years ago, as the world cooled towards the Pleistocene era. The tundra and the windswept grasslands of the steppe were inhabited by lemmings and voles, rodents small enough to retreat in winter to shelter under the snow, not only from the icy wind but also from most aerial hunters. Northern stoats and weasels small enough to hunt them there have a winter-long larder mostly inaccessible to other predators. For the same reason, fossils of stoats that lived during the glacial periods were smaller than those dated to warmer periods (Marciszak & Socha, 2014). Today, most larger northern mammals migrate south during the coldest months, but stoats and their prey remain, protected by their chilly duvet.

Winter nests built by voles and lemmings offer essential refuges to their slim, agile hunters. Nests are linked by runways, accessible by occasional tunnels leading to the surface. The rodents can still access the grass without having to dig for it, and the stoats and weasels are small enough to hunt through the runways and raid the winter nests. Their short legs and long bodies make running through the confined spaces natural and easy, but their thin fur and skinny frames make them very vulnerable to the cold. So they take over the best and safest nests for bed as well as board, lining them with vole fur and stocking their larders with frozen dinners.

The main problem with this beneficial arrangement is that vole and lemming populations are notoriously unstable. Every 3 or 4 years, small mustelids are surrounded by more rodents than they can eat. Nestling stoats born in those years are well fed, and many of them survive to disperse in search of their own territories. The next season, hunting is poor to impossible, and few if any young stoats survive to independence. For a female stoat, the difference between a good and a bad season defines her reproductive success, her legacy to future generations. The most successful females are those that maximise the number of living young she can produce in a good year, and minimise her losses in a bad year.

Delayed Implantation Delayed implantation is a common feature among the mustelids, which suggests it was an ancestral character that many of the living species have inherited. Its main effect is to separate the mating and birthing seasons by an interval longer than the gestation period. The issue is one of timing. The start of the mating season is tightly synchronised by the increasing day lengths of spring. The optimum time for the adult males to search for mates is determined by the best environmental conditions for travelling, which for stoats is in spring or early summer. But the young conceived then take 2–3 months to grow to full independence, by which time, in northern climates, the winter is fast approaching. Delayed implantation offers a way to escape this mismatch in seasons by postponing the start of the short active gestation until conditions favour the survival chances of the dispersing young.

The stoats transported to New Zealand inherited this fixed reproductive cycle with very little variation. Males have two periods of intense sexual activity: spring (September–October in New Zealand), when the adult females that have failed to rear a litter first come into oestrus; and early summer (October–November), when the precocious juvenile females and their mothers become receptive. The onset of spermatogenesis and the implantation of the blastocysts both tend to be later in the far south of New Zealand (44–45°S) by a short period (about 10 days) corresponding to the lag in the date at which the southern regions reach the critical day length that sets off these processes in mid to late winter.

Proof of how daylength controls the date of implantation can be provided by manipulating the lighting over the cages of captive stoats. Simulated summer photoperiods applied in winter significantly advanced the onset of oestrus and conception date by 2 months earlier than in untreated females, so artificially extending the breeding season of captive stoats (O'Connor et al., 2004).

Hence, most North Island litters are born from late September to early October, and most South Island ones from mid to late October. During their short receptive period, oestrous females develop large pink swellings in their vulval area. Ovulation is stimulated only by mating, so unmated females can remain on heat for weeks. After ovulation, cross sections of the ovaries show large cellular structures, the corpora lutea, which fill the spaces vacated by the released ova, one to one (Fig. 4.5). Corpora lutea therefore provide certain confirmation of mating, and ovulation rate can be counted by anyone with the histological machinery needed to see them. No more litters can be conceived for another 12 months, even if the forest is swarming with rodents.

Immediately after mating, the fertilised eggs develop for 2 weeks into blastocysts, then float free in the uterus throughout summer and autumn until late winter. The corpora lutea of delay maintain the blastocysts by producing levels of ovarian hormones too low to stimulate implantation. The oestrus period is brief, and ovulation is induced, but the males are efficient at finding receptive females. In the large National Parks survey collections, only two of 451 females we collected between December and July inclusive carried no corpora lutea. The proverbial English description of a particularly persistent human suitor as 'a bit of a stoat' is clearly a great compliment.

Fecundity Because the corpora lutea mark the ovulation rate one to one, counting them in serial sections gives an estimate of fecundity, that is, the number of ova fertilised, the *potential* litter size. The females in this condition are fertilised, but not yet investing any serious energy in the litter. Implantation is the process that sets off an active pregnancy lasting 28 days, and once it is triggered in August, the birth date of the litter is fixed. But the *actual* litter size—number of kits born—is controlled quite differently.

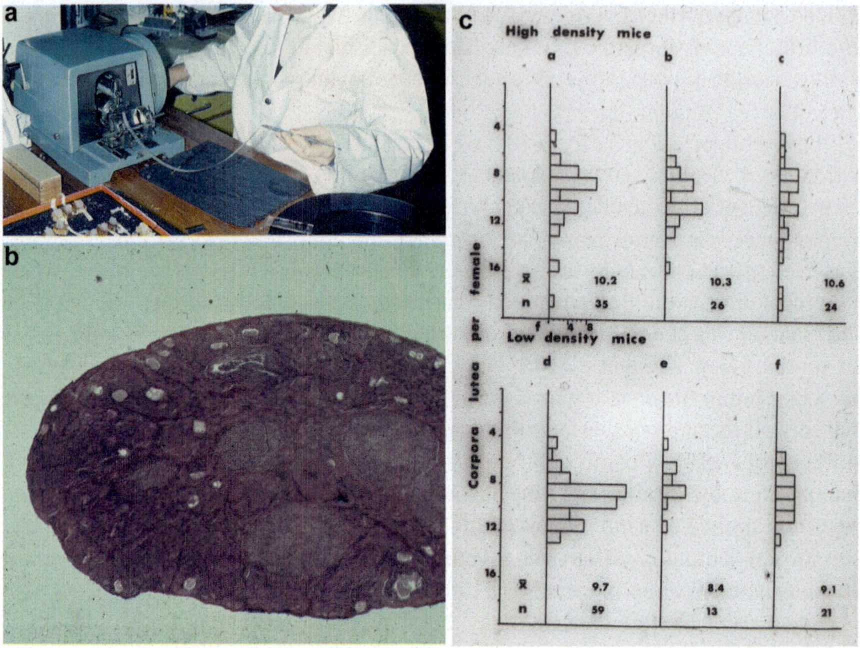

Fig. 4.5 (**a**) Serial sectioning of stoat ovaries embedded in wax blocks. A rotary microtome runs off a ribbon of sections at 15 μm. The ribbon can then be cut it by eye into eight equal portions 160–290 μm apart; two adjacent sections were mounted and stained, 16 to a slide. Corpora lutea appeared in 2–4 pairs of sections, so could be counted by plotting their presence or absence through the series. (Photo by Author). (**b**) Corpora lutea are large solid bodies that can still be distinguished even in poor material. (Photo by Author). (**c**) Frequency distribution of corpora lutea counts in relation to abundance indies of mice. (From King et al, 2003).

Because fertilisation and implantation are held so far apart by delayed implantation, stoats are among the very few animals in the temperate world to conceive and produce their litters in different years. At the time of mating, female stoats depending on unstable food supplies, such as a fluctuating population of voles or mice, cannot know what environmental conditions will be like by the time the kits are born. If rodents are then scarce, a female producing more kits than she could feed would waste her own resources. If it turns out to be a bumper year for rodents, she cannot add an extra litter, so would miss an opportunity for maximum reproductive success. In other words, productivity (number of young born) is not related to fecundity (ovulation rate). Living stoats still carry the genes that have successfully determined another way to adjust their fecundity to the unknown future chances of bearing large (or any) litters of young.

Across all the different New Zealand habitats we sampled, the average ovulation rate in 439 females was a relatively constant 9.7 ova, range 3–20, regardless of environmental conditions at the time of mating. It costs a young female very little to maintain the corpora lutea of delay while her own body is still growing, so the blastocysts remain merely viable, floating free in the uterus, throughout the period of delay.

But active pregnancy is very energy-demanding, and after implantation the females' food requirements suddenly accelerate. If she is well fed, she has a good chance of implanting all or most of the blastocysts she already carries, and of retaining the embryos through to viable birth weights after an active gestation of 4 weeks. In a bad season for rodents, when she is herself in poor condition, she fails to implant some blastocysts, or aborts some embryos. She cannot respond to a good rodent year by producing an extra litter; she can only minimise the mortality rate within the potential litters whose numbers (now reduced to an average of 8.8 embryos, range 6–13) were fixed months ago.

In a really bad year, females can resorb some or even all of the embryos, or lose the entire litter at the nestling stage. We could detect such cases because the annual reproductive cycle of stoats is strongly synchronised and predictable from date and latitude of capture. So a female caught during the breeding season with fresh corpora lutea at a time when all the other females examined in that month were actively pregnant or lactating, must have lost the potential litter she had been carrying over winter and had been re-mated early, so was already fertilised for the next season 12 months ahead. We found eleven such individuals, all collected from places where we knew that mice were scarce that year.

The advantage of delayed implantation is that, as conditions improve early next spring, female stoats are *already* set to produce a litter at the optimum moment for them, in numbers adjusted to the rodent resources of the time. She does not have to find a mate or wait for the young to develop and grow from fertilisation to independence. If she has dispersed alone to an offshore island like Motutapu (Sect. 7.4), or a fenced sanctuary like Orokonui (Sect. 13.2), she can start a new population unaided.

Precocious Young Females These are all substantial advantages for adult animals with a good chance of surviving through more than one birthing season, like martens or badgers (the stoat's larger relatives which also have delayed implantation). But small body size and short lives go together, which is a disadvantage for young stoats who have to survive nearly a year before they can produce their first litter. First-year mortality rates among stoats are generally high, and in some years, catastrophic (Sect. 11.4). Worse, if young females first mated at 12 months old, as do young males, delayed implantation would ensure that their litters could not be born until they were nearly 2 years old. Not many of them live that long (Sect. 5.2).

Sexual bimaturism means that infant female stoats pass puberty whilst still in the nest, indeed while they are still blind and nearly hairless. The adult males are attracted to their mother's den, helped by the scent and squeaks of the young. If the mother is in residence, any male strong enough to mate with her (a process sometimes requiring vigorous negotiation) has a good chance of entering the den.

All the young of that age have a short brown mane of fur, covering a patch of skin glands whose scent is different in males and females. The male picks up each young in turn, immediately dropping the males, and mating with the females. Even though they are still so tiny, they are fully fertile. Adult males can serve the juvenile females without damaging them, as illustrated by one 17-day-old female kit observed in a captive population in Russia. She was only 112 mm long, weighing 18 g (13% of the weight of the mother, and 6% of that of the adult male), blind, deaf, helpless, and almost immobile. She was mated for 1 min, and in the following season produced 13 kits and fed them successfully (Ternovsky, 1983).

This scenario does not amount to an animal version of rape. As soon as the young females catch the male's scent, they become very excited, and squeak loudly to attract his attention (Müller, 1970). It is in their interest to be fertilised as nestlings. That means they leave already carrying next year's litter, a huge advantage for dispersal in years and places when males are scarce. Their brief encounter with the adult males induces ovulation and immediate fertilisation, which after 2 weeks passes into the same period of delayed implantation as in the adults. By the time the adult male leaves, all the female young and their mother are fertilised.

The enormous advantage of this arrangement for stoat researchers is that, for most of the year, every female stoat of any age is fertilised. Counts of corpora lutea can be made from almost all females collected, for some 10 months the year. This system helps to explain why stoats are so resistant to control. Routine trapping usually removes more males than females (Sect. 4.7), but cannot reduce the reproductive potential of the population unless it is particularly intensive.

The combination of these three unusual reproductive characteristics gives stoats a unique capability of matching their annual productivity to a fluctuating food resource (King et al., 2003). When small rodents, their favourite prey, respond to good food supplies by staging a sudden population irruption, female stoats are already carrying large litters of *potential* young (Sect. 5.3). During the inevitable decline of the rodent population, female stoats reduce their litter size to the number they can feed. In a really bad crash year, the adult females abandon any attempt to breed (Purdey et al., 2004), and survive on alternative prey until the next season. New Zealand is the best place to obtain the large samples required to document these processes (Sect. 1.4), but they probably apply to stoats everywhere.

4.6 Stoat Age Classes

Juvenile survival is the key factor determining how stoat numbers vary from year to year (Table 5.1). The process of trying to understand stoat population dynamics has to start with knowing what controls the overwinter mortality of the young, and how long the adults live. That in turn requires a reliable method of age determination. As in English weasels (Sect. 2.4), the two most informative bones needed for this purpose are the skull and baculum, so these bones were routinely extracted and cleaned during the first-stage processing of stoat carcases.

Simple Morphology The skull of a stoat is not exactly a larger copy of that of a weasel, but it is near enough to allow the DSB (date-skull-baculum) method based on known-aged weasels (Chap. 2) to be applied as a first attempt. The bacula of stoats and weasels are different in shape, but equally responsive to the seasonal bathing in testosterone that produces the distinct proximal knob distinguishing adults. For classifying our material, the main problem at first was that there was no reference material of known age. Yet most aspects of stoat biology are affected by age, so it could not be ignored.

A long search throughout the 1970s failed to find any reliably age-related cues. One of them, a small knob on the femur called the lateral suprasesamoid tubercle, was claimed to develop only in adults, so we cleaned the femurs of our stoats as well as their skulls. The supposed correlation was there, but not reliable enough to be used for age determination. On the other hand, cleaning the femurs turned out to have been worthwhile anyway, for other reasons (Sect. 4.2). Eventually I published a review paper describing the trials, in the hope of helping others to avoid wasting time doing the same (King, 1991b).

The obvious place to start was to apply the DSB method. The changes with age in the shape of the skull of weasels and stoats are similar, so it was reasonable to assume that the age-related characters of the two species should have much in common. The method is reliable for as long as the skulls of the young of the year are clearly distinguishable from those of adults caught in the same month. How long might that difference remain distinct? A trial I ran at DSIR Ecology Division laid out a very large series of clean female skulls in twelve monthly groups according to date of death. Within each group, five independent observers were asked to separate the young of the year from the adults. The 'accuracy' of their results was unknown, but the number of errors could not be less than the number of 3:2 disagreements among the five observers.

Seasonal Limits For stoats collected in summer (December to February inclusive), fewer than 4% of each sample caused that level of disagreement. For every other month of the year, the five observers disagreed on at least 20% of each monthly group. The conclusion was that the young (they cannot be called juveniles, since the females are sexually mature) of either sex could be reliably recognised up to the last month of summer (Fig. 4.6). From then until late winter, subadult males could be distinguished from adults from the baculum, if it was available. All females caught after reaching 5 months old, and males caught after 11 months old, were classified as nominal adults.

Wild stoats can live for 3 years or more, so the inability to subdivide the adults was frustrating. But an opportunity to assemble a set of known-aged stoat skulls did eventually appear (Sect. 5.1). Meanwhile, the analyses of the main collection proceeded using the minimal reliable age categories that could then be recognised.

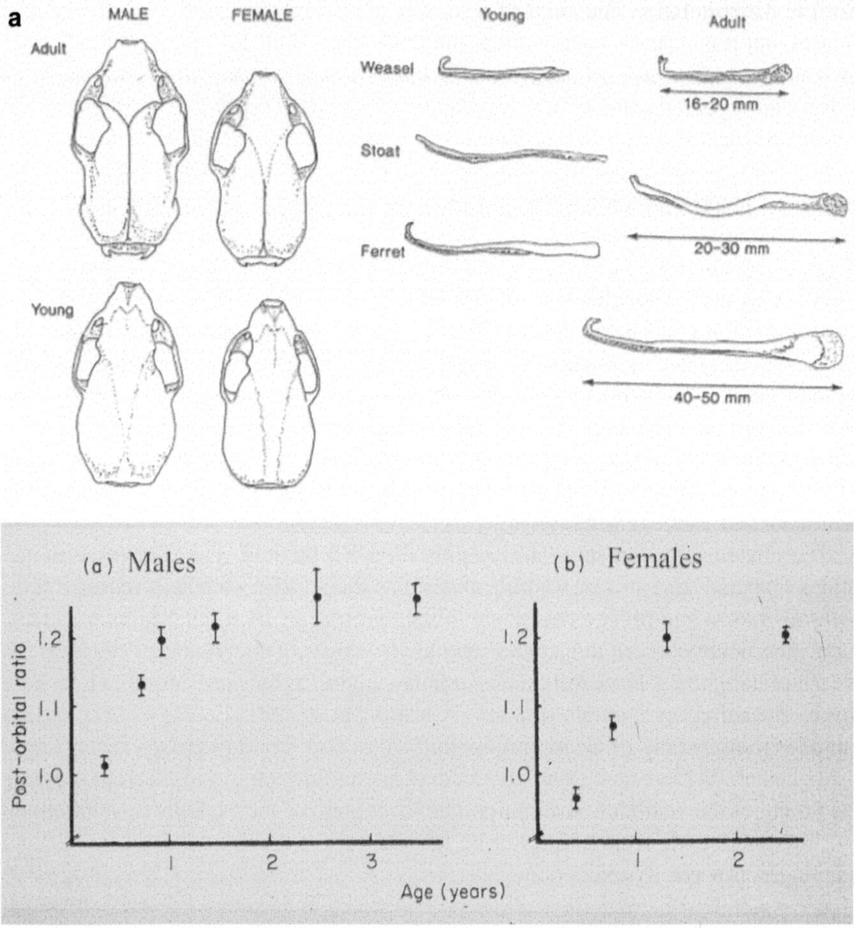

Fig. 4.6 (**a**) Skulls of male and female stoats, young and adult, plus the bacula of young and adult male stoats, weasels and ferrets. (Drawings by Consie Powell). (**b**) Development of the post-orbital ratio of stoats with age. (From King, 1991b)

4.7 Seasonal Variation in Sampling

Over the long term, controlled collections made with widely spaced kill traps can give a lot of information useful to population studies. Kill traps don't have to be inspected daily, so can be spread over a wide area, but removing some individuals can affect the density, age structure, social relationships, and dynamics of the rest. So for research on populations, systematic kill-trapping data corrected for effort are needed. On the other hand, obtaining large samples of carcases from populations already subject to trapping for fur, or for control to protect prey or game populations, can easily provide important information on, say, reproductive cycles and age structure that is impossible to obtain any other way. Some patterns in our survey results were too obvious to be derailed by uneven sampling.

Sex Ratio Among the total of 1599 stoats collected from around New Zealand through the year, there was a very clear seasonal cycle in the capture rate per month. In November (spring), only 74 stoats were collected, all but four of them adults. From December to February inclusive (summer), the capture rate increased steadily as the annual cohort of young of both sexes appeared in the traps. Total captures of all ages climbed from 148 to 185, then 224 per month. From March to May (autumn) the totals steeply declined (129, 119, 131 per month), then dropped to below 100 a month over winter until spring (September). A slight blip in October (110) could perhaps be related to the renewed activity related to the mating season. Females are especially hard to catch after the end of delayed implantation, when they tend to retreat to a well-stocked den to await the birth of their litter. Hence, the sex ratio of captures from August to November was always >70% males, compared with <70% over the rest of the year.

Not much else could be concluded at the time from such raw figures with minimal ecological background, except that the probability of catching a stoat of any age is not constant through the year, especially not for females. In summer the newly independent young of both sexes are still learning about traps so are easily caught, but many of them disappear without trace over winter. The really critical data, including parallel estimates of predators and prey, became possible only during the next stage of the work (Chap. 5).

Trap Spacing The result that the park authorities were most anxious to see was some estimate of how much damage stoats were doing to the park's wildlife. Some of the individual park rangers went to extraordinary extra lengths to run as many traps as possible, on top of all their other ordinary work. For example, Ray Slater had frequently to drive along the Milford road from Te Anau to the Main Divide as part of his routine job (Fig. 4.1). For more than 18 months (November 1972–June 1974, a period of low stoat numbers between seedfalls) he took the time to stop and check a trap line along a 40 km stretch of the Eglinton Valley (King, 1980).

In his 29 traps, Ray caught 123 stoats, and regarded every one as a contribution to protecting his favourite birds, the mōhua (*Mohoua ochrocephala*). His unstated assumptions were that trapping can reduce the number of stoats hunting birds in the valley, and that predation by stoats controls the numbers of mōhua. He and his colleagues were disappointed to be told that there was no way to test those assumptions from his trapping data at the time, but in fact, his efforts became important to a later analysis.

In an experiment asking what would be the results of setting traps at 100 m, 200 m, 400 m and 800 m intervals, we added Ray's results to represent an extreme spacing of >2 km. He helped us confirm that most females escape widely spaced traps. Females are the most important individuals to remove, so these results confirmed other thinking about how to catch more of them.

Ray may well also have helped the birds more than he knew then. Since the late 1980s, the Eglinton Valley has become an important centre for field research.

Long-running programmes building on earlier work in the valley (Sect. 14.2.2) have clearly confirmed the benefits of stoat control in at least temporarily reducing the effect of stoat predation on hole-nesting birds like mōhua and kākā, and ground-foraging native bats.

4.8 From Data to Information

Pre-computer Technology Such a huge mass of data collected by our survey took years to analyse. Everything we knew about each stoat was assembled onto its own individual edge-punched card. In the days before computers, these were the quickest way to sort and tabulate records. Every punch hole around the four edges of the card denoted one character, say male or female. Using a special pair of pliers, the outside edge of the hole representing the character of interest was punched out. To pull out all the males, one would hold the whole stack of cards in one hand, and with the other push a mounted needle through the hole marked 'Males'. A gentle shake caused all the cards on which that hole had been punched to fall out, leaving only the undamaged cards representing the females or unknowns.

Repeating the process identified all the females, leaving only the unknowns The number of cards describing definite males and females could then be counted. To subdivide the males into, say, adults and young, the same process would be repeated again for the holes representing age classes. This unwieldy-sounding system, the only one available, worked well for small samples, and allowed an outlying result to be traced back immediately to everything known about the animal concerned from the details entered on the cards that had fallen out. But not for a stack of 1599 cards.

Fortunately, by early 1977 DSIR had a mainframe computer which could read its own style of punch cards, and do extensive statistics on them, so the data were transferred onto a different set of cards in the form the computer could read. All the original classifications had to be converted into digital categories, a process inviting multiple risks of errors in translation. That in itself took months, as one of us read from the edge-punched cards and the other entered the digital codes into the machine-readable cards. All the consequent analyses were run by a contractor with the required skills in IT and statistics. The report he provided became the basis of a series of papers edited with skill and patience by Cleveland Duval (King & Moody, 1982).

References

Erlinge, S. (1987). Why do European stoats *Mustela erminea* not follow Bergmann's rule? *Holarctic Ecology, 10*, 33–39.

Johnstone, A. C., & Garvey, P. M. (2023). To boldly go: Methods to quantify personality in mustelids. *Animal Behaviour, 202*, 139–147.

King, C. M. (1980). Field experiments on the trapping of stoats (*Mustela erminea*). *New Zealand Journal of Zoology, 7*, 261–266.

King, C. M. (1981). The effects of two types of steel traps upon captured stoats (*Mustela erminea*). *Journal of Zoology (London), 195*, 553–554.

King, C. M. (1989). The advantages and disadvantages of small size to weasels, *Mustela* species. In J. L. Gittleman (Ed.), *Carnivore behaviour, ecology and evolution* (Vol. 1, pp. 302–334). Cornell University Press.

King, C. M. (1991a). Body size—Prey size relationships in European stoats *Mustela erminea*: A test case. *Holarctic Ecology, 14*(3), 173–185.

King, C. M. (1991b). A review of age determination methods for the stoat *Mustela erminea*. *Mammal Review, 21*(1), 31–49.

King, C. M., & Moody, J. E. (1982). The biology of the stoat (*Mustela erminea*) in the national parks of New Zealand. *New Zealand Journal of Zoology, 9*, 49–144.

King, C. M., & Powell, R. A. (2007). *The natural history of weasels and stoats: Ecology, behavior and management* (2nd ed.). Oxford University Press.

King, C. M. (2021). Survival of femur fractures in wild stoats (Mustela erminea). *Animal Welfare, 30*, 461–466.

King, C. M., White, P. C. L., Purdey, D., & Lawrence, B. (2003). Matching productivity to resource availability in a small predator, the stoat (*Mustela erminea*). *Canadian Journal of Zoology, 81*, 662–669.

King, C. M., Norbury, G., & Veale, A. J. (2017). Small mustelids in New Zealand: Invasion ecology in a different world. In D. W. Macdonald, C. Newman, & L. A. Harrington (Eds.), *The biology and conservation of musteloids* (pp. 257–277). Oxford University Press.

Little, L., King, C. M., & O'Donnell, C. F. J. (2017). Behaviour of stoats (*Mustela erminea*) raiding the nests of rock wrens (*Xenicus gilviventris*) in alpine New Zealand. *Notornis, 64*, 124–135.

Marciszak, A., & Socha, P. (2014). Stoat *Mustela erminea* Linnaeus, 1758 and weasel *Mustela nivalis* Linnaeus, 1766 in palaeoecological analysis: A case study of Bisnik Cave. *Quaternary International, 339-340*, 258e265.

McAulay, J. R., Monks, J. M., Wilson, D. J., & Seddon, P. J. (2021). Individual specialists within a generalist niche: Variable diet of stoats and implications for conservation. *New Zealand Journal of Ecology, 45*(2), 3443.

McDonald, R. A., & Harris, S. (2002). Population biology of stoats *Mustela erminea* and weasels *Mustela nivalis* on game estates in Great Britain. *Journal of Applied Ecology, 39*, 793–805.

McDonald, R. A., Webbon, C., & Harris, S. (2000). The diet of stoats (*Mustela erminea*) and weasels (*Mustela nivalis*) in Great Britain. *Journal of Zoology (London), 252*, 363–371. https://zslpublications-onlinelibrary-wiley-com.ezproxy.waikato.ac.nz/doi/epdf/10.1111/j.1469-7998.2000.tb00631.x

Müller, H. (1970). Beiträge zur Biologie des Hermelins, *Mustela erminea* Linné, 1758. *Saugetierkundliche Mitteilungen, 18*(4), 293–380.

Murphy, E. C., Gillies, C., Maddigan, F., McMurtrie, P., Edge, K.-A., Rohan, R., & Clapperton, B. K. (2016). Flexibility of diet of stoats on Fiordland islands, New Zealand. *New Zealand Journal of Ecology, 40*, 114–120.

Noonan, M. J., Johnson, P. J., Kitchener, A. C., Harrington, L. A., Newman, C., & Macdonald, D. W. (2016). Sexual size dimorphism in musteloids: An anomalous allometric pattern is explained by feeding ecology. *Ecology and Evolution, 6*, 8495–8501.

O'Connor, C. E., Turner, J., Scobie, S., & Duckworth, J. (2004). Stoat reproductive biology. *Science for Conservation, 268*, 24.

Powell, R. A. (1982). Evolution of black-tipped tails in weasels: Predator confusion. *The American Naturalist, 119*, 126–131.

Powell, R. A., & King, C. M. (1997). Variation in body size, sexual dimorphism and age-specific survival in stoats, *Mustela erminea* (Mammalia: Carnivora), with fluctuating food supplies. *Biological Journal of the Linnean Society, 62*(2), 165–194.

Purdey, D. C., King, C. M., & Lawrence, B. (2004). Age structure, dispersion and diet of a population of stoats (*Mustela erminea*) in southern Fiordland during the decline phase of the beechmast cycle. *New Zealand Journal of Zoology, 31*, 205–225.

Segal, A. N. (1975). Postnatal growth, metabolism, and thermoregulation in the stoat. *Ékologia, 1*, 38–44.

Ternovsky, D. V. (1983). The biology of reproduction and development of the stoat *Mustela erminea* (Carnivora, Mustelidae). *Zoologischesky Zhurnal, 62*(7), 1097–1105.

Vattiato, G., Binny, R. N., Davidson, S. J., Byrom, A. E., Anderson, D. P., Plank, M. J., Carpenter, J. K., & Alex James, A. (2023). Detectability of ten invasive mammal pests in New Zealand: A synthesis of spatial detection parameters. *New Zealand Journal of Ecology, 47*, 3552.

Warburton, B., et al. (2008). Traps for killing stoats (*Mustela erminea*): Improving welfare performance. *Animal Welfare, 17*, 111–116.

White, P. C. L., & King, C. M. (2006). Predation on native birds in New Zealand beech forests: The role of functional relationships between stoats and rodents. *Ibis, 148*, 765–771.

Part III

Part III Different Habitats, Different Lives

The Beechmast Cycle, 1972–81

5

Abstract

The masting cycle of the southern beeches, Nothofagaceae, is synchronised by climate cues. Mass seedfalls provide resource pulses at 3–5-year intervals which drive cascading shifts in distribution, abundance and population dynamics of the whole beech forest community, from native invertebrates and seed-eating birds to introduced predators. Forest rodents (wild house mice (*Mus musculus*) and ship rats (*Rattus rattus*)), continue breeding over the winter after a heavy seedfall, reaching high abundances by late spring and early summer followed by a sudden crash in autumn and winter. Juvenile mortality in stoats (*Mustela erminea*) is much lower than usual, leading to a large cohort of newly independent first year stoats searching for prey just as the rodents are declining. Predation on forest birds is most intense during the post-seedfall summer, slackening as young stoats starve. These predictable events enable conservation managers to organise expensive and timely stoat and rat control programmes, with good effects for populations of native birds in the managed areas. An independent research programme used the same cues to catch, mark and release 134 stoats during the seedfall year of 1979–80. Fourteen of them were retrieved to confirm that baculum weight and cementum annuli are reliably linked with known age.

Keywords

Masting cycles · Resource pulsing · Predation risk · Age determination of stoats

The Orongorongo data (Chap. 3), few though they were, pointed to the importance of simultaneous monitoring both seedfall and rodents in at least some of the areas supplying carcases. So as an extension of the general National Parks Stoat Survey (Chap. 4), we set up further regular field observations to continue documenting the interlinked population changes of stoats and rodents in the three most accessible and

C. M. King, *Stoat in the Dock*, Fascinating Life Sciences,
https://doi.org/10.1007/978-981-96-4562-6_5

best organised collecting areas (the Eglinton and Hollyford valleys in northern Fiordland National Park, and Craigieburn Forest Park, in Canterbury). We covered two major seedfall events, 1976 and 1979, but even larger ones have been recorded since.

The Beech Masting Cycle To monitor seedfall, we set up lines of Beveridge seed collecting trays, with known collecting areas of 0.28 m^2, alongside quarterly 36-station rodent trap lines modelled on those in the beech forests above the Orongorongo Valley. We already knew that the good mast season of 1976 in South Island beech forests would predictably stimulate a mouse irruption, followed by a damaging plague of hungry stoats. But we couldn't show exactly how this cycle worked, or when was the best moment to disrupt it, or when was the moment of greatest danger for birds. So, the next step was to document this sequence of events in greater detail.

The irregular production of seed by the southern beeches, and the following irruptions of rats and mice, had long been known from simple observation. Even before European settlement, when the only rodent species present was the kiore or Polynesian rat *Rattus exulans*, Māori hunters knew to expect a bumper harvest of kiore in their traps after a seedfall (Best, 1942). In 1956, a report of an early survey in southern Fiordland, written by pioneer ecologist Thane Riney, included a simple statement summarising all the major elements of the beechmast cycle, correctly deduced without relevant field evidence by intuition alone (Riney et al., 1959).

What was not known was exactly *how* the sequence worked. How could the cascading reactions of forest fauna to a heavy seedfall produce such damaging numbers of hungry predators so quickly? If the sequence of population processes involved could be better understood, it might be possible to turn these observations into an effective management strategy timed to have the greatest impact on predator control at the most critical moment.

Organising Support At a memorable meeting of Fiordland National Park Board towards the end of the survey programme, I explained to the board members why it would be important to dissect the process in greater detail through the next heavy masting cycle. One board member was a staunch believer in the kill-'em-all view, and he was convinced that no research not focussed on finding more and better ways to kill stoats was worth funding. 'Why are you wasting time looking at seeds and mice, when you should be just killing more stoats?' he demanded. Fortunately, other members of the board could indeed see the point, especially the Chief Ranger, Allan Cragg, supported by key members of the National Parks Authority in Wellington, Bing Lucas and Ben Thorpe. Without their support, our programme might have stopped right there.

A simple monitoring method , requires that the traps be set out evenly, baited and checked daily, and operated in the same way and for the same number of days per session regularly all the year round. The theory is that, if the trapping technique has been absolutely consistent, and if the catch data are corrected for variation in trapping effort, then changes in the number of animals killed probably reflect real changes in the numbers of animals available to be killed. Likewise, if the living stoats of all ages and both sexes are caught at the same rate (or at least, that differences between them are constant), then changes in the proportions of animals of each age and sex caught should reflect real changes in the structure of the population observed.

It is important to spell out these rather obvious prerequisites, because much depends on them, and they are not always as true as one might expect. For example, if traps are baited, it could be argued that when more stoats are caught it is because they are more hungry, not because there are more of them. If that happens, then changes in the actual numbers of stoats become confused with changes in their willingness to enter traps ("trapability"). Capture-per-unit-effort indices cannot distinguish changes in numbers from changes in trapability. So there are many precautions to remember about the perils of making conclusions about living populations of stoats from analyses of carcases. Concerns about variable trapability later proved valid (Sect. 11.2).

The Park Board approved a new contract extending the general, national-level 1972–76 stoat collections through to 1981 in the three best-monitored areas in which standardised field methods could be imposed (Eglinton, Hollyford and Craigieburn). Together they supplied twelve seasons' beech seedfall, 1138 captures of rodents, 905 carcases and 162 scats of stoats. The seedfall status of stoat birth years could be deduced for all the stoats examined except those born before 1964 at Craigieburn and before 1969 in Eglinton/Hollyford, whose data were removed from the analyses. The results documented the consequences of the beechmast cycle for stoats through two consecutive heavy seedfalls, in 1976–77 and 1979–80, in ways that exceeded all expectations (King, 1983), and has been confirmed in much greater detail since (Wilson & Lee, 2010).

Clearly, in order to know to what extent stoats kill birds, and to have some chance of predicting where and how to concentrate control efforts, managers need to know not only what the stoats are eating (Sect. 4.3), but what determines the numbers of stoats from year to year, and how long they live. To answer that question, it was first necessary to acquire some means of estimating age-specific fertility and mortality rates of wild stoats.

5.1 Stoat Birth Certificates

At that time, the programme was still reluctantly confined to very basic age determination by simple observation of the skull and baculum (Fig. 4.6). However, others were developing a much more precise method, focussed on layers in the cementum at the roots of the canine teeth made visible by sectioning. In hibernating animals such as hedgehogs, these are often clearly annual, marking a seasonal interruption in growth rate over winter. But for stoats, the assumption that the lines are annual had first to be proven from known-aged animals, and we had none.

Collecting Known-Aged Animals In the summer of 1978–79, the southern beech trees in Fiordland National Park put on a huge flowering event, which led to predictions of a heavy seedfall from March to June of 1979. The anticipated stoat irruption over the 1979–80 summer offered an opportunity to obtain the known-aged reference material acquired from local wild stoats. I designed a project to live-capture, mark and release the large sample of the young stoats expected to appear in December, January and February, in the hope of recapturing enough of them over the next few years to provide a reference group with known capture records.

The Fiordland National Park Board granted a contract for an intensive addition to our stoat data collections through that summer, and employed a trapper, Chris Macmillan, to operate two lines of live traps in the Eglinton and Hollyford Valleys. Over a short but hectic period between late November 1979 and the end of January 1980, he caught, tagged and released 134 stoats (113 young of the year, many still distinctly smaller than adults, and with no signs of breeding) plus 19 adults and two unknowns. Such a massive capture rate was far higher than expected, and every stoat had to be anaesthetised for tagging, so he often did not finish his daily trap round until the late evening (King & McMillan, 1982).

Over the following 18 months, 14 marked stoats, from 107 last seen alive, were recovered, and another eight were kept in captivity for 15 months. The canine teeth of these 22 animals were sectioned, and in the 20 that returned clear results, the number of growth rings in the dental cementum corresponded exactly with known age in the 12 marked as young (Fig. 5.1). In the eight marked as adults, the number of lines was equal to or greater than the minimum number of years of known age (Grue & King, 1984).

The number of marked animals returned was fewer than expected, because (as became clear only much later) the project was misunderstood by local people to be

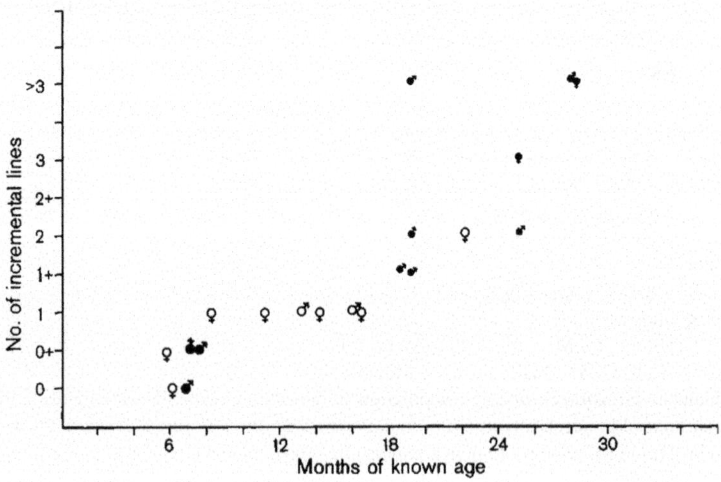

Fig. 5.1 Cementum annuli in canine teeth of known-aged stoats. (From Grue & King, 1984)

importing and releasing new stoats in the forest, not merely marking those already there. Even those who knew what we were doing disagreed that it could ever be permissible to release a captured stoat alive. It was a sharp early lesson in the need (never before considered) to involve the local community in any field project. Like most scientists, and to my later regret, I sometimes got things wrong on the way to getting then right. But in the end the results produced enough positive agreements, and no contrary cases, to support the assumption that cementum lines in stoats are indeed annual.

5.2 How Long Can Stoats Live?

Weasels and stoats are small, and on average they are short-lived, because their survival prospects are closely related to the wildly variable numbers of small rodents during their first year. But average age tells nothing about the longevity of individuals. Some far outlive their contemporaries, but many more die young. So for our wider project, we needed not only a means of distinguishing annual age classes of adults, but also to collect large enough samples to find the rare, older individuals.

Stoats can have only one birthing season a year, and that one tightly synchronised across all habitats by daylength (Sect. 4.5), so the entire national cohort of young stoats is born within a month in early summer. Successive cohorts are easily distinguishable for at least 6 months, and the young males are clearly recognizable until shortly before the next season (Fig. 5.2). In all nine studies surveyed (King & Powell, 2007: 271), in seven different countries, the first year class (the cohort born in the most recent birthing season) always comprised more than half the sample.

The beechmast cycle caused massive variation in breeding success for female stoats and in the survival prospects of their young. The first year cohort ranged from 15% in poor mouse years to 92% during a mouse peak. Much later, in the Borland Valley during the crash year of 2000–2001, the whole potential cohort of young was wiped out. All 65 adult females examined in December/January by Purdey et al. (2004) had lost their litters, reducing the proportion of young of the year to zero.

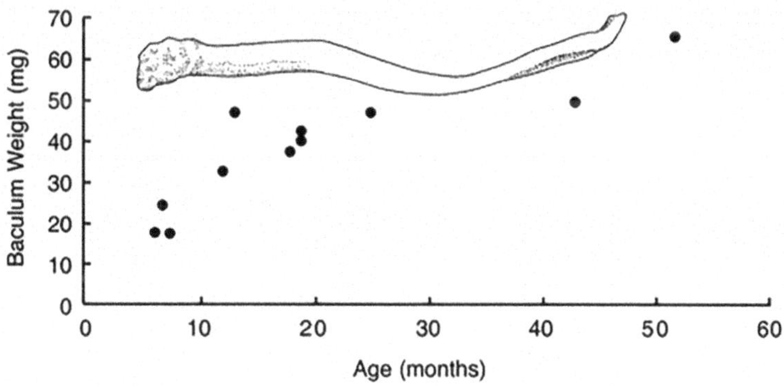

Fig. 5.2 The weight of the stoat baculum plotted against known or part-known age increases for years, perhaps for life. (Age data from Grue & King, 1984. Drawings by Consie Powell)

Likewise, in the post-peak summer of 1991–1992, no young stoats appeared among the 37 adult stoats caught (Murphy & Dowding, 1995).

By contrast, the huge cohorts of young produced in bumper seasons can be followed through several successive years. Normally, the second year class comprises the smaller number of young of the previous year that had survived past their first birthday, and successive year classes drop away smoothly to oldest identifiable. But the beechmast cycle can produce a quite different picture. For example, the very large cohort of young (almost 90% of 183) born during the mouse peak of 1976–77 was followed by a very small cohort of young representing minimal stoat breeding success during the mouse crash year of 1977–78. But in that year, the 1-year-olds born in 1976 were still almost as numerous as the young of the 1977 season. In 1971, a year of huge abundance of stoats throughout the country (Sect. 3.1), the record 1971 cohort was still evident in the 1973–1974 catch as a bulge of 2–3 year-olds.

Effects of Trapping on Age Structure If trapping really can affect a stoat population, the older stoats should disappear sooner from a trapped than from an undisturbed area. For example, in three samples taken over several years in New Zealand beech forests, the mean age of stoats caught in the early years of a programme was significantly higher than in later years, declining from 19 months to 10 months in both Eglinton and Hollyford valleys, and 16 months to 10 months at Craigieburn (Powell & King, 1997). At Pureora, a nonbeech forest (Chap. 6), the proportion of stoats older than 1 year old declined over 5 years of removal sampling, from 52% of 21 collected in summer and autumn of 1983 to 27% of 22 collected in the same two seasons of the 4 years from 1984 to 1987 (King et al., 1996).

So, in severe crash years, when the total output of young is zero, a whole annual cohort may be deleted from the population age structure. Natural examples of that extreme result have been documented twice, both times in beech forests during the hungry season following a mouse population crash. For short-lived animals, loss of a complete generation over a large enough area could have a drastic effect on their numbers. That would be exactly the result most sincerely desired by conservationists worried about the ethics of lethal trapping and poisons. Would it not be useful for stoat control programmes to ask how such a natural effect could be engineered deliberately?

The very oldest age class that could be identified from the cementum annuli in the canine teeth comprised stoats 7 years old or more, and these appeared only in one small study in Denmark ($n = 28$) and two much larger kill-trapped samples from Russia ($n = 616$) and New Zealand ($n = 223$). Previous protection from trapping helps; one killed on the road to Milford Sound had lived to almost 5 years old despite a broken leg (Fig. 4.3), and the first catches on Resolution Island included several 6 year olds (A. Veale, Personal communication). Only a very few individuals lived to be 8–10 years old, so only a very large sample could find them.

In theory, being born in a favourable year should translate into a permanent advantage—a condition known as 'a silver spoon effect'. This idea does apply to some longer-lived species, but for stoats, the silver spoon effect is completely reversed. The good food supplies in mouse peak years following a seedfall certainly allowed young stoats to grow well for their first few months, so in our data, many more larger than average first-year stoats were caught in than in non-seedfall years (Sect. 4.4). On the other hand, those newly independent young were swiftly pitched into huge competition for food as the mouse and rat populations diminished in autumn. That is, they became independent just as the rodents were already disappearing. Life for this cohort of stoats swung rapidly from feast to famine within a few months, with drastic consequences for their chances of surviving to breed in their turn.

Age-specific mortality rates among the extra-numerous cohorts of young of the year born in seedfall years was very high in all three of our study areas ($q_0 = 0.92$, 0.91, 0.92 versus 0.73, 0.55, 0.60 in other years) (Wittmer et al., 2007). So proportionately more stoats born in *poor* mouse years lived longer lives (mean age at death for males 16.1 months, females 11.5 months) than did those born during a post-seedfall mouse peak (males 11.0 months, females 10.9 months) (Powell & King, 1997). Among the 40 males of more than 3 years old identified in that study, only seven were the larger individuals that had been presumably well fed as nestlings as they grew up while mice were abundant. The other 33 were the smaller-sized males that had been born in a non-seedfall year. The same applied, less obviously, to females: 17 of 30 females born in a mouse-poor year lived to over 3 years old, compared with the 13 presumably better-fed females that died younger. No plausible explanation of this surprising but certainly significant difference has yet appeared.

The national parks survey data set is by far the largest collection of reliably age-classified stoat carcasses known, and provides the basic information required to study the effects of unstable prey populations on successive cohorts of a predator. It documents the significant age-specific differences in body size, survivorship and fecundity that are essential for constructing population models (Sect. 11.4). Critics might argue (and some did) that deducing such information from dead animals would be statistically invalid, but live stoats do not often lend themselves to the necessary landscape-scale field tracking of marked individuals or large samples, and the critical data on skull size and fecundity can be acquired only from carcasses. So, despite the various questions arising from how our field data were collected (Sect. 4.1), and other concerns about replication and absence of objective experimental design, we took the view that an exploratory survey such as this one can still be useful on the scale required and within broad limits, and lead to consistent new insights into stoat biology not attainable any other way. After all, any project can get further by being more or less accurate if not precise than by being precise but inaccurate.

5.3 The Sequence of Events Through the Masting Cycle

A good seedfall of southern beech is preceded by a mass flowering event in early summer, when the canopy of the beech forest is painted a glowing red by the huge numbers of tiny scarlet male beech flowers. On a still morning, a red mist of beech pollen drifts across a mountain valley, and settles as red dust on the surface of a calm lake. By autumn (March–May) the little prickly seed cases appear, and in due course they release a small package of triangular seeds. Occasionally the rate of seed-set is low, meaning that many seed cases contain no kernels, so the promise of a good flowering season is not always fulfilled. But if most of these seeds are sound, the forest floor becomes strewn with a bonanza of extra food for invertebrates, mice, rats, seed-eating birds and (indirectly) for stoats, followed by predictable population irruptions of each in turn.

The advantage of masting is that seed predators cannot exhaust a large crop of extra food delivered all at once. Some of the seeds are likely to survive if the crop is large enough to satiate all potential seed-eaters, provided all the trees drop their seeds at the same time. And, indeed, that does work, so in spring after a huge mast year, the forest floor is carpeted with tiny beech seedlings. Moreover, the southern beeches are not the only masting species in New Zealand. In the other principal forest type, native rimu and other podocarp species have their own masting cycles, precipitating similar events (Ruscoe et al., 2004). Alpine tussocks also have mast seasons which fuel predator populations above the tree line (Kelly et al., 2008). In all three habitats, masting events set off a trophic cascade of interactions affecting the entire forest community—from spent flowers through invertebrates to rodents and birds to the local top predators.

Winter Breeding of Mice Wild house mice are the only small rodents resident in New Zealand's cool beech forests (Walker et al., 2019), and they are smaller than the voles and lemmings which the stoats' ancestors had evolved to hunt. Ship rats are more sensitive to cooler temperatures than mice, so were present in low-elevation Hollyford, scarce in mid elevation Eglinton, and absent in high-elevation Craigieburn until global warming allowed then to invade long after our sampling finished (Harris et al., 2022). Ship rats are more common than mice in podocarp forests (Chap. 6), but mast years have similar effects on them too (Harper, 2005).

Mice normally come into breeding condition in spring, and after a short gestation can produce a litter of six in a few weeks. The adult females can produce several litters a year, and the young females can breed in the year of their birth. As the weather cools, rodent breeding slows, and usually stops over winter. After a heavy seedfall, events take a quite different turn (Fig. 5.3).

When the forest floor is strewn with extra food, female mice continue to produce young, who rapidly bear their own young throughout the winter. By the time mice hit the peak of their irruption in late spring, most are of breeding age. As the age structure of the resident population of mice shifts steadily upwards, the older mice

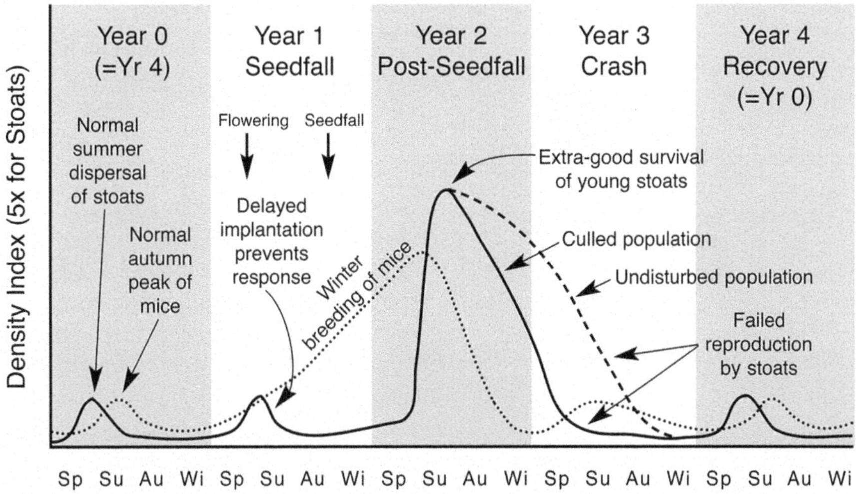

Fig. 5.3 The sequence of events through an average 4-year beech mast cycle. (Redrawn by Roger Powell from King, 1983)

stop breeding, and by early summer the population recruitment rate falls to zero (Fitzgerald et al., 2004).

Stoats respond to a sudden irruption of prey by varying the critical parameters that control age-specific fertility and mortality rates. Cohorts are counted from October–September, so fecundity and fertility fall in successive years. The most important variable is the variation in the survival of juveniles from implantation to independence.

Well-fed female stoats produce most or even all of the young they have been carrying as blastocysts since last spring (Sect. 4.5). They easily find plenty of mice to feed their nestlings. As one might expect, more mice are eaten per stoat as the density of mice increases (Jones et al., 2011). In the early summer after a heavy seedfall, breeding stoats and their kits are surrounded by abundant easy prey. No wonder that most of the young stoats born then survive to independence. They dominate the extra-large cohorts that remain distinguishable in annual population structures for years. By mid summer, this huge cohort of hungry young stoats disperses throughout the forest, fiercely competing for mice just as the numbers of mice begin a steep decline. Mouse irruptions inevitably collapse after young mice cease to join the population, whether stoats are present or not.

So over the autumn and winter following a post-seedfall summer, the advantage of a seed-year cohort goes into reverse as mice disappear rapidly, followed by stoats. By spring of the next year, adult female stoats are in bad condition, and most or all of their litters conceived during the good times are lost. They still mate and conceive the usual number of young, which go into delayed implantation as usual. None of the young conceived then can be born until the next year, but many of the adults can survive on alternative prey until next season. The population rates of increase during

Table 5.1 Population dynamics of stoats through a typical post-masting sequence

	Seedfall year	Post-seedfall year	Stoat crash year	Transition year(s)
Stoat parameters	Mice increasing	Stoat peak; mouse peak then declining	Mice gone	Between stoat crash and mast years
Fecundity (CL/female)	9.7 (set before pulse, to be born during peak)	10.2 (to be born in crash year)	8.9 (to be born next year)	9.2 (to be born next year)
Female body condition	Good	Excellent	Poor	Moderate?
Fertility (kits born)	1.9 ± 0.01	5.0 ± 0.01 (after high intra-uterine survival)	1.72 ± 0.01 (after high intra-uterine mortality)	1.45 ± 0.01
Survival age 0–1	0.2 ± 0.03	0.50 ± 0.03	0.08 ± 0.02 (after high juvenile mortality)	0.40 ± 0.03
Survival age 1–2	0.60 ± 0.03	0.40 ± 0.03	0.30 ± 0.05	0.20 ± 0.03
Survival >2	0.50 ± 0.01	0.50 ± 0.01	0.50 ± 0.01	0.50 ± 0.01

The key factor is food-controlled variation in survival through age class 0–1

each phase of the cycle averaged (+ SD) for Seed years 2.87 + 0.04, for Post-seed years 0.39 + 0.00, for Crash years 0.91 + 0.00 and for Transition years 0.99 + 0.05 (Wittmer et al., 2007). The net result summarised in Table 5.1 explains the extraordinary capacity of stoats to match their productivity to an unstable supply of resources (compiled by Roger Powell from King et al., 2003).

The feast-or-famine conditions of life in pure beech forests is difficult for other predators, including weasels, which are too small and short-lived to survive or emigrate during the mouse-poor periods between mast years, but stoats live longer and have access to more alternative prey. The ancestors of stoats were already adapted to very similar conditions imposed by the fluctuations of lemmings and voles in the Arctic. This pre-adaptation of stoats to unstable food supplies made them the worst possible predator to deliberately introduce into New Zealand.

5.4 The Danger Period for Birds

Back in 1976 the data showed only that birds comprised an average of some 43% or so of the general stoat diet, varying greatly between locations and years. But stoats have a list of profitable prey, and mammals are usually higher on it than are birds. The larger, more nutritious rats, possums or rabbits offer the large proportions of meat that stoats can easily strip off the bones, while avoiding the indigestible fur. Hence, in habitats supporting enough mammalian prey, stoats will tend to prefer them while available, and put up with having to spit out bird bones and feathers only when necessary.

Rodents Can Buffer Birds, or Vice Versa Northern hemisphere experience shows that, as a rodent irruption provides stoats with enough to eat of their favourite mammalian prey, they tend to leave the birds alone. For example, in Wytham Wood, the Edward Grey Institute found that fewer tit (Paridae) nest boxes were raided by weasels when the woodland voles and European wood mice were abundant (Sect. 2.2). On open farmland in southern England, populations of field voles varied between years from near zero to around 200 voles/ha. The diets of weasels surrounded by voles included <10% birds, but that proportion more than tripled after the vole population collapsed (Tapper, 1979). In New Zealand, the story is not quite so simple.

The key question determining how many birds are taken by stoats is whether there are enough alternative sources of nutrition available to distract stoats' attention from birds. That is because a stoat can get more energy from 50 g of rat or rabbit meat free of indigestible fur and bones than from a 50 g bird. At Okarito, a wildlife reserve on the South Island's west coast, stoats ate more rats and fewer birds between the spring of 2002 and the winter of 2004, during a time when the number of ship rats available (body weight averaging around 120–160 g) was rising (Murphy et al., 2008). In that case, the birds were indeed buffered from the stoats' attentions, at least temporarily, because stoats preferred to eat rats whenever there were enough of them.

But if those nutritious mammals suddenly disappear, as was observed in Pureora Forest Park after a successful rat-poisoning operations, stoats deprived of rats turned to birds (Murphy et al., 1998). Likewise, in the Mackenzie, rabbit haemorrhagic disease (RHD) in 1997 removed most of the stoats' favourite prey. The arrival of RHD coincided with a sudden increase in predation on banded dotterel nests, from a normal average of 22% to 50–75% (Norbury et al., 2002). The same effect predictably followed a successful 1080 operation against rabbits in the same area (Murphy et al., 2004). In those cases there were not enough mammalian prey to buffer the birds against stoats.

Beech forests are now pretty marginal habitats for stoats in most years. The abundant native ground fauna of the nineteenth century are all gone. House mice average only 15–20 g in body weight, so they offer much less substantial food sources for stoats than do rats, although both can be abundant after a seedfall. Stoats respond rapidly to an irruption of mice, and for a short while they do eat a lot of mice, but mice alone are simply too small to support a lot of stoats for long, and mice very rarely get to numbers high enough to distract stoats from eating birds (White & King, 2006).

Only briefly, during an exceptional mouse peak, has a buffering effect typical of the northern hemisphere been observed in New Zealand. In the Borland Valley during the summer of 1999/2000, after a huge seedfall when mouse captures got up to 62% trap response (more than twice the usual post seedfall peak), only 5% of the 87 stoat guts examined contained birds, and 87% had mice. Of the stoats collected during the following mouse crash year, 2000/2001, 48% of 63 guts contained birds, and 10% mice (Purdey et al., 2004). The rest of the time, a higher number of mice merely supports more stoats to eat birds.

Rodent Crash Years Beech forests offer few other mammalian prey, so when the mice and rats are few, the stoats have few alternatives. Stoats might eat more birds then, but when rodents disappear, so—after a while—do stoats. Over a mouse-poor summer when stoat numbers are low, 10 stoats destroying 100 nests with eggs and chicks each cannot remove nearly as many birds as 50 stoats destroying 100 nests each in a mouse peak year. Conversely, after a moderate fall of beechmast, the inevitable rodent irruption boosts the normal summer peak in stoat numbers (the key time of year for survival of young birds) by up to five times higher than usual. Against all reasonable expectations, about the same proportion of birds appear in stoat guts in all summers, meaning that peak numbers of stoats all eating the usual numbers of birds per head inevitably added up to greater risk for birds (King, 1983). In most post-seedfall summers, a moderate increase in number of mice merely supports more stoats to eat birds. Higher numbers of rats also account for more birds, often more than stoats do (Carpenter et al., 2025).

The clearest illustration of this danger period was provided by the DSIR Ecology Division's team then based in Nelson Lakes National Park, in the northern South Island (Wilson et al., 1998). They monitored the masting cycle and its influence on seedfall, mice and stoats using the same methods as described above. They provided independent confirmation of the consequences of rodent and stoat irruptions through the sequence. Their main aim was to document what effect introduced predators were having on the decline of kākā (*Nestor meridionalis*) populations. Their results are described in Sect. 14.2.7.

References

Best, E. (1942). *Forest Lore of the Maori*. Dominion Museum.

Carpenter, J. K., Verry, A. J. F., Mitchell, K. J., Trought, K., Monks, A., Innes, J., Griffiths, J., Harvey, M., & Wilmshurst, J. M. (2025). DNA metabarcoding of stomach contents shows predation impacts of black rats on Island birds after a resource pulse are driven by numerical not functional response. *Biological Invasions, 27*, 57. https://doi.org/10.1007/s10530-024-03522-z

Fitzgerald, B. M., Efford, M. G., & Karl, B. J. (2004). Breeding of house mice and the mast seeding of southern beeches in the Orongorongo Valley, New Zealand. *New Zealand Journal of Zoology, 31*, 167–184.

Grue, H. E., & King, C. M. (1984). Evaluation of age criteria in New Zealand stoats (*Mustela erminea*) of known age. *New Zealand Journal of Zoology, 11*, 437–443.

Harper, G. A. (2005). Heavy rimu (*Dacrydium cupressinum*) mast seeding and rat (*Rattus* spp.) population eruptions on Stewart Island/Rakiura. *New Zealand Journal of Zoology, 32*, 155–162.

Harris, H. A. L., Kelly, D., Innes, J., & Allen, R. B. (2022). Invasive species and thermal squeeze: Distribution of two invasive predators and drivers of ship rat (*Rattus rattus*) invasion in mid-elevation *Fuscospora* forest. *Biological Invasions*. https://doi.org/10.1007/s10530-022-02789-4

Jones, C., Pech, R., Forrester, G., King, C. M., & Murphy, E. C. (2011). Functional responses of an invasive top predator *Mustela erminea* to invasive meso-predators *Rattus rattus* and *Mus musculus*, in New Zealand forests. *Wildlife Research, 38*(2), 131–140. https://doi.org/10.1071/wr10137

Kelly, D., Turnbull, M. H., Pharis, R. P., & Sarfati, M. S. (2008). Mast seeding, predator satiation, and temperature cues in *Chionochloa* (Poaceae). *Population Ecology, 50*, 343–355.

King, C. M. (1983). The relationships between beech (*Nothofagus* sp) seedfall and populations of mice (*Mus musculus*),and the demographic and dietary responses of stoats (*Mustela erminea*), in three New Zealand forests. *Journal of Animal Ecology, 52*, 141–166.

King, C. M., & McMillan, C. D. (1982). Population structure and dispersal of peak-year cohorts of stoats (*Mustela erminea*) in two New Zealand forests, with especial reference to control. *New Zealand Journal of Ecology, 5*, 59–66.

King, C. M., & Powell, R. A. (2007). *The natural history of weasels and stoats: Ecology, behavior and management* (2nd ed.). Oxford University Press.

King, C. M., Flux, M., Innes, J. G., & Fitzgerald, B. M. (1996). Population biology of small mammals in Pureora Forest Park: 1. Carnivores (*Mustela erminea, M. furo, M. nivalis,* and *Felis catus*). *New Zealand Journal of Ecology, 20*, 241–251.

King, C. M., White, P. C. L., Purdey, D., & Lawrence, B. (2003). Matching productivity to resource availability in a small predator, the stoat (*Mustela erminea*). *Canadian Journal of Zoology, 81*, 662–669.

Murphy, E. C., & Dowding, J. E. (1995). Ecology of the stoat in *Nothofagus* forest: Home range, habitat use and diet at different stages of the beech mast cycle. *New Zealand Journal of Ecology, 19*(2), 97–109.

Murphy, E. C., Clapperton, B. K., Bradfield, P. M. F., & Speed, H. J. (1998). Effects of rat-poisoning operations on abundance and diet of mustelids in New Zealand podocarp forests. *New Zealand Journal of Zoology, 25*, 315–328.

Murphy, E. C., Keedwell, R. J., Brown, K. P., & Westbrooke, I. (2004). Diet of mammalian predators in braided river beds in the central South Island, New Zealand. *Wildlife Research, 31*(6), 631–638. https://doi.org/10.1071/wr03033

Murphy, E., Maddigan, F., Edwards, B., & Clapperton, K. (2008). Diet of stoats at Okarito Kiwi Sanctuary, South Westland, New Zealand. *New Zealand Journal of Ecology, 32*(1), 41–45.

Norbury, G., Heyward, R., & Parkes, J. (2002). Short-term ecological effects of rabbit haemorrhagic disease in the short-tussock grasslands of the South Island, New Zealand. *Wildlife Research, 29*, 599–604.

Powell, R. A., & King, C. M. (1997). Variation in body size, sexual dimorphism and age-specific survival in stoats, *Mustela erminea* (Mammalia: Carnivora), with fluctuating food supplies. *Biological Journal of the Linnean Society, 62*(2), 165–194.

Purdey, D. C., King, C. M., & Lawrence, B. (2004). Age structure, dispersion and diet of a population of stoats (*Mustela erminea*) in southern Fiordland during the decline phase of the beechmast cycle. *New Zealand Journal of Zoology, 31*, 205–225.

Riney, T., Watson, J. S., Bassett, C., Turbott, E. G., & Howard, W. E. (1959). Lake Monk expedition, an ecological study in Southern Fiordland. *DSIR Bulletin, 135*, 1–75.

Ruscoe, W. A., Wilson, D. J., McElrea, L., McElrea, G., & Richardson, S. J. (2004). A house mouse (*Mus musculus*) population eruption in response to a heavy rimu (*Dacrydium cupressinum*) seedfall in southern New Zealand. *New Zealand Journal of Ecology, 28*, 259–265.

Tapper, S. (1979). The effect of fluctuating vole numbers (*Microtus agrestis*) on a population of weasels (*Mustela nivalis*) on farmland. *Journal of Animal Ecology, 48*, 603–617.

Walker, S., Kemp, J. R., Elliott, G. P., Mosen, C. C., & Innes, J. G. (2019). Spatial patterns and drivers of invasive rodent dynamics in New Zealand forests. *Biological Invasions, 21*, 1627–1642.

White, P. C. L., & King, C. M. (2006). Predation on native birds in New Zealand beech forests: The role of functional relationships between stoats and rodents. *Ibis, 148*, 765–771.

Wilson, D. J., & Lee, W. G. (2010). Primary and secondary resource pulses in an alpine ecosystem: Snow tussock grass (*Chionochloa* spp.) flowering and house mouse (*Mus musculus*) populations in New Zealand. *Wildlife Research, 37*(2), 89–103. https://doi.org/10.1071/wr09118

Wilson, P. R., Karl, B. J., Toft, R. J., Beggs, J. R., & Taylor, R. H. (1998). The role of introduced predators and competitors in the decline of kaka (*Nestor meridionalis*) populations in New Zealand. *Biological Conservation, 83*, 175–185.

Wittmer, H. U., Powell, R. A., & King, C. M. (2007). Understanding contributions of cohort effects to growth rates of fluctuating populations. *Journal of Animal Ecology, 76*(5), 946–956. https://doi.org/10.1111/j.1365-2656.2007.01274.x

6 Podocarp-Broadleaved Forest

Abstract

The New Zealand Forest Service once managed Pureora Forest for intensive harvest of native timber trees and replanting with *Pinus radiata,* until all logging was stopped in 1978. They commissioned a 5 year survey of small mammals in Pureora 1982–87 describing the effects of past logging/replanting on the prospects of pest control and post-logging restoration. The same field protocols previously used in beech forests were applied again, with very different results. By contrast with cooler beech forests, where mice are abundant but only after a seedfall year (Chap.5), in richer and warmer forests such as Pureora the dominant rodents every year are rats. Among a total of 2740 captures, a 5-year survey collected 1793 ship rats (*Rattus rattus*), and 522 mice (*Mus musculus*), but only 57 stoats (*Mustela erminea*), 16 weasels (*M. nivalis*) and 11 ferrets (*M. furo*). The average stoat reached the age of 15 months, after a first-year survival rate of about 24%. But the proportion of first-years declined from 52% to 27% over the 5 years, showing that trapping stoats in non-beech forests can exceed their natural replacement rate. Pureora is now part of the Department of Conservation estate.

Keywords

Pureora Forest Park · Logging history · Rodent distribution by forest type · Impact of trapping on stoats

Background The National Parks Stoat Survey and its spinoff projects in southern beech forests all finished field work in 1981 (Chaps. 4 and 5). For the next decade, the New Zealand science establishment offered no permanent professional opportunities for a specialist carnivore biologist, other than contract work for outsiders who had the experience and the time. This chapter describes one such contract initiated by the New Zealand Forest Service in 1982.

C. M. King, *Stoat in the Dock*, Fascinating Life Sciences,
https://doi.org/10.1007/978-981-96-4562-6_6

Fig. 6.1 Pureora Forest Park supports patches of untouched tall podocarp forest, a completely different habitat for stoats than the southern beech forests described in Chap. 5. SCION 8042505 (1979). (Photo by J. Johns)

Pureora Forest, west of Lake Taupo on the volcanic plateau of the central North Island, is totally different from the southern beech forests (Fig. 6.1). The present community of native flora and fauna is a non-beech forest that has been formed, destroyed and re-formed by millennia of dramatic volcanic events and natural forest successions, centuries of human exploitation both by traditional hunting and by mechanized logging, and decades of change wrought by introduced pests and by their opponents, the organised, official management action devoted to pest control and restoration work (King et al., 2015).

6.1 The Contract

In 1982, the New Zealand Forest Service decided to set up a 5-year programme to monitor small mammals in Pureora Forest Park. The idea was to make controlled comparisons of the abundance of introduced predators in two contrasting forest types, using exactly the same methods in both. The populations of small mammals in North Island podocarp forest—a completely different environment from southern beech forest—were expected to offer a contrasting view of predator community dynamics compared with those of the South Island.

It turned out that concentrating on these two study areas was an unplanned but lucky choice. Later analyses summarized the complex range of New Zealand forest rodent dynamics into six classes based on decades of tracking tunnel data described by Walker et al. (2019). The six types range from 'irruptive', which have, as in Fiordland, low median but synchronous high abundance of mice, to 'continuously ratty', which have, as in Pureora, high median rat and low unsynchronised mouse abundances year-round.

So, quite by chance, these early field projects happened to describe the background needed to compare the opposite ends of the spectrum of rodent population dynamics, and the consequences for stoat populations, in completely different forest ecosystems. All the same, it is important to remember that our results apply only to these relatively simple forest types, not to New Zealand ecosystems in general.

The study areas included samples of three different non-beech forests affected by increasing degrees of logging, different dates of re-planting with *Pinus radiata*, rapid growth of pines at the expense of ground cover, and other human disturbance. In all these environments, the rodents and their predators were doing different things.

At the end of November 1982, a new set of trap lines was established, and then operated quarterly in February, May, August and November of 1983–87. Collaboration with colleagues John Innes, Marie Daugherty and Meg Flux was essential to complete the field work on rodents, and with the local rangers running the stoat lines.

Over the 5 years, this systematic survey in Pureora Forest Park, using a strictly standardised regime of traps, collected samples of nearly all the small mammal species resident in the forest, including 522 mice (*Mus musculus*) and 1793 ship rats (*Rattus rattus*), but only 57 stoats (*Mustela erminea*), 16 weasels (*M. nivalis*) and 11 ferrets (*M. furo*) (King et al., 1996b).

Ship rats lived "wall-to-wall" throughout all indigenous forest types, logged or not, but they were scarce or absent in exotic forest. Of 179 sites used throughout this programme at which Fenn traps were set for rats and stoats, only 7 failed to catch at least one ship rat in 5 years: at 14 sites, 20 or more rats were caught. Analyses of this abundant material led by Mark Kimberley provided us with baseline information on relative numbers, distribution, population structure and breeding biology of both species in relation to habitat.

Long-term changes in the abundance of ship rats are usually estimated in terms of a trapping index (% trap success, corrected for occupied traps). This index varies between years, seasons and habitats, and is strongly affected by trap type, food supplies and the success of the previous breeding season for rats and stoats. The general average density of ship rats in North Island forest is about 3.9/ha, compared with 0.003/ha for stoats (Innes et al., 2010).

Impact of Trapping on Stoats In 24,272 trap nights using Fenn traps at Pureora, set over 5 years and equally good at catching all three small mustelids, only the stoats provided enough data for analysis, and they proved to be very different from the stoats collected by identical methods from South Island beech forests. They were most often caught in the grass along roadsides and in older exotic forests, where they were more likely to find mice and rabbits, but they never exceeded the numbers typical of southern non-seeding years.

The same field and lab methods previously described (Chap. 5) showed that average age of stoats at capture was 15 months, and their average annual mortality rate from independence to 1 year old was about 76%. The proportion of stoats in the sample aged over a year old declined over the 5 years of the collections, from 52% of 21 collected in summer and autumn of 1983 to only 27% of 22 collected in the same two seasons of 1984–87 (King et al., 1996a). These results suggest that the natural density and productivity of stoats in non-beech forest are both relatively low, and a determined trapping campaign has some chance of exceeding their natural mortality rate and temporarily reducing their local density.

6.2 Dramatic Changes at Pureora

Pureora had been a standard logging village established in the late 1940s to provide the complex mill and housing facilities serving the last timber companies working in native forest (King et al., 2015).

The upheaval that shook Pureora Forest and all its residents is not only dramatic as a human story, but also recent and exceptionally well recorded. Within a few years after logging stopped in 1978, the mills were closed, the timber companies and most of their employees moved away, and the remaining forest was totally protected. Pureora became the focus of worldwide concern for the survival of threatened New Zealand forest fauna generally, especially a rare endemic bird the North Island kōkako (*Callaeas wilsoni*), and on how they can be best preserved into the future. In many ways, the history of this one area is a microcosm of the conservation history of New Zealand generally.

Why did all this happen in Pureora in the mid twentieth century, as opposed to any other time and place? Part of the answer is that the environmental movement, focussed on the last few populations of endemic forest fauna, became a force to be reckoned with only in the 1970s. They were able to force the New Zealand Forest Service to stop the logging, and to commission urgent research on predator control, only in the 1980s. The outcome is an internationally recognized case history of the costs and benefits of conservation versus logging of old-growth native forest (King, 2023).

Had any of these previous events happened any sooner, it would have been too late to commission the internationally significant research that saved the kōkako (Innes et al., 1999). This rare bird, that most Pureora residents had never seen, was already extinct in many other more accessible parts of its formerly very wide range, but is now thriving (Sect. 14.2.5). The last valuable tall native forests of the central North Island (Fig. 6.1) would have been clear-felled and converted to farms or pine plantations (Fig. 6.2), just as had those that had already gone from nearly two million hectares of more accessible land elsewhere.

Ironically, in the context of this book, Pureora provides an example of a critically important conservation success in an environment where the predatory impact of ship rats was much more significant than that of stoats.

Fig. 6.2 The aftermath of clearfelling indigenous forest, burning the remains and replanting with exotic species in the ashes, according to standard practice in 1977. Three small *Pinus radiata* pine seedlings are visible in the foreground. (Crown Copyright, Department of Conservation Te Papa Atawhai. Photographer unknown)

6.3 The Creation of the Department of Conservation (DOC)

The New Zealand Forest Service (NZFS) had for decades been responsible for both production forestry and for conservation of native timber. These two largely contradictory functions sometimes caused conflicts within the ranks. Seldom mentioned was the advantage that, at the same time, some commercial revenues could also quietly be diverted to support conservation purposes, for example, by funding the 5-year survey of small mammals at Pureora. Without that discreet but invaluable support, thanks to Andy Leigh, the Pureora project would have been impossible.

Nevertheless, for political more than practical reasons, described at length by King et al. (2015), at the end of March 1987 the formerly powerful New Zealand Forest Service was dis-established. Those of its former staff not concerned with production forestry were transferred to a new entity, the Department of Conservation (DOC), along with others from the New Zealand Wildlife Service (NZWS), the Historic Places Trust and the National Parks Authority.

NZFS transferred to DOC some 3 million hectares of State Forest land, plus the 1.8 million ha State Forest Parks network, plus the control and management of native wildlife and game birds, recreational fishing, and wild game animals in New Zealand. DOC was therefore to become responsible for most of the natural lands and forests of the Crown, including 10 Recreational hunting areas totalling 290,000 ha, 14 forest sanctuaries (16,000 ha) and 48 ecological areas (160,000 ha) (New Zealand Forest Service, 1986).

Research in any of these areas is now done only under permit , so DOC has been deeply involved in all forms of field studies ever since.

References

Innes, J. G., Hay, R., Flux, I., Bradfield, P., Speed, H., & Jansen, P. (1999). Successful recovery of North Island kokako *Callaeas cinerea wilsoni* populations, by adaptive management. *Biological Conservation, 87*, 201–214.

Innes, J., Kelly, D., Overton, J. M., & Gillies, C. (2010). Predation and other factors currently limiting New Zealand forest birds. *New Zealand Journal of Ecology, 34*(1), 86–114.

King, C. M. (2023). The costs and benefits of conservation versus logging of old-growth native forest: A case history. *Ecological Economics, 204*, 107632.

King, C. M., Flux, M., Innes, J. G., & Fitzgerald, B. M. (1996a). Population biology of small mammals in Pureora Forest Park: 1. Carnivores (*Mustela erminea, M. furo, M. nivalis,* and *Felis catus*). *New Zealand Journal of Ecology, 20*, 241–251.

King, C. M., Innes, J. G., Flux, M., Kimberley, M. O., Leathwick, J. R., & Williams, D. S. (1996b). Distribution and abundance of small mammals in relation to habitat in Pureora Forest Park. *New Zealand Journal of Ecology, 20*, 215–240.

King, C. M., Gaukrodger, J. D., & Ritchie, N. A. (2015). *The drama of conservation: A history of Pureora Forest, New Zealand.* Springer/New Zealand Department of Conservation.

New Zealand Forest Service. (1986). *Ministerial Brief for Minister of Conservation*. New Zealand Forest Service.

Walker, S., Kemp, J. R., Elliott, G. P., Mosen, C. C., & Innes, J. G. (2019). Spatial patterns and drivers of invasive rodent dynamics in New Zealand forests. *Biological Invasions, 21*, 1627–1642.

Part IV

Part IV Advances in Technology

Island Eradications 7

Abstract

Islands can offer safe refuge for vulnerable native species, so New Zealand is fortunate to have >250 islands managed as reserves. Stoats can swim for at least 2 h non-stop at steady 1 km/h, potentially longer if assisted by a current or a floating log. Stoats have got onto some 100 islands, but have never reached the most important ones. Stoats have been eradicated from 15 islands of >100 ha, and those located >5 km from shore are usually safe. Resolution and Secretary Islands are rugged enough to make trapping expensive and difficult, and both are well within swimming range of stoats. Stoats resident on Resolution island breed fast and avoid traps, but can be suppressed enough to allow some reintroductions of native species. Mop-up operations with toxins cannot be used where native species are at risk. Any incursion on a stoat-free island already repopulated by sensitive native fauna demands an immediate and expensive emergency response. Intense genetic monitoring of stoats on Waiheke Island since 2016 continuously follows the movements and effective trap avoidance behaviour of stoats resident there. New Zealand experience is being applied to eradication of stoats on the UK Orkney Islands.

Keywords

Island invasions · Swimming stoats · Eradication histories · Historical invasions · Predator-free refuges · Island eradications · Stoat swimming range · Island invasive species eradications database · Island incursions database · Cost of incursion responses · Genetic monitoring

Introduction New Zealand is an archipelago of 9800 islands (collectively accounting for 1.2% of New Zealand's total land area), of which 1500 are larger than half a hectare. Of the 735 inshore and offshore islands of >1 ha, 250+ are managed as

C. M. King, *Stoat in the Dock*, Fascinating Life Sciences,
https://doi.org/10.1007/978-981-96-4562-6_7

reserves; 158 have never had any mammal invasions (Towns et al., 2012). Still, at *least* 167 separate island populations of native birds are known to have gone from the main and offshore islands since 1000 CE (Holdaway et al., 2001).

These isolated populations, once protected from intruders by their marine defences, are desperately vulnerable to invasion, and many of those lost in the past were local endemics. Often, the damage done by invaders is under-estimated, or even unknown, until native species return after their habitat is restored. For example, a previously unknown species of snipe (*Coenocorypha* undescribed sp.) recolonised Campbell Island (11,300 ha) from tiny Jacquemart (2000 ha) only after the eradication of Norway rats in 2001 (Miskelly & Fraser, 2006).

Well-managed offshore islands offer effective lifeboats for native species unable to tolerate the introduced predators on the mainland. Both The Department of Conservation (DOC) and its predecessor the New Zealand Wildlife Service (NZWS) have invested decades of effort and millions of dollars in attempts to sweep as many islands as possible free of aliens, in order to create safe homes to which vulnerable species can be transferred. Better still, the remaining islands still hosting different combinations of alien mammals—say, with rats but not stoats, or vice versa—can offer useful comparisons helping to identify the damage done by one or the other separately.

7.1 Historic Record

Only a handful of the hundreds of known lost populations of native fauna could have been affected by stoats, and a further 112 were never in contact with stoats, for two reasons. First, stoats arrived late, after the most vulnerable mainland birds had already been removed by Polynesian and European hunters, rats, cats, dogs, and various drastic habitat changes. Second, stoats have never reached the most important offshore islands on which many endemic bird populations have survived the disturbances on the mainland. Nevertheless, stoats have quickly become a very real threat to some of our most important island refuges.

More than 7500 stoats were liberated on Otago pastures damaged by rabbits over the decade from 1884, and rapidly spread across the mainland (Sect. 1.2). Stoats are excellent swimmers, so are or have been present on some 100 islands, and eradicating them became a high priority task as soon as the technology became available (Veale et al., 2012). Their distribution is consistent with the principles of island biogeography: the larger and closer to the mainland, the greater the chances of a swimmer arriving (Russell & Clout, 2004). Stoats have been removed, at least temporarily, from 15 islands over 100 ha in size in Aotearoa (Fig. 7.1). Most of the long list of other island eradications have involved rats, not stoats.

A recent survey by Peter Corson (2024) of the records available in the Department of Conservation's Island Incursion Database (Clout & Russell, 2006), and the Island Invasive Species Eradications (islandconservation.org), both constantly updated, found at least 46 incursions by stoats in the 10 years 2010–2019. Of these, 42 were

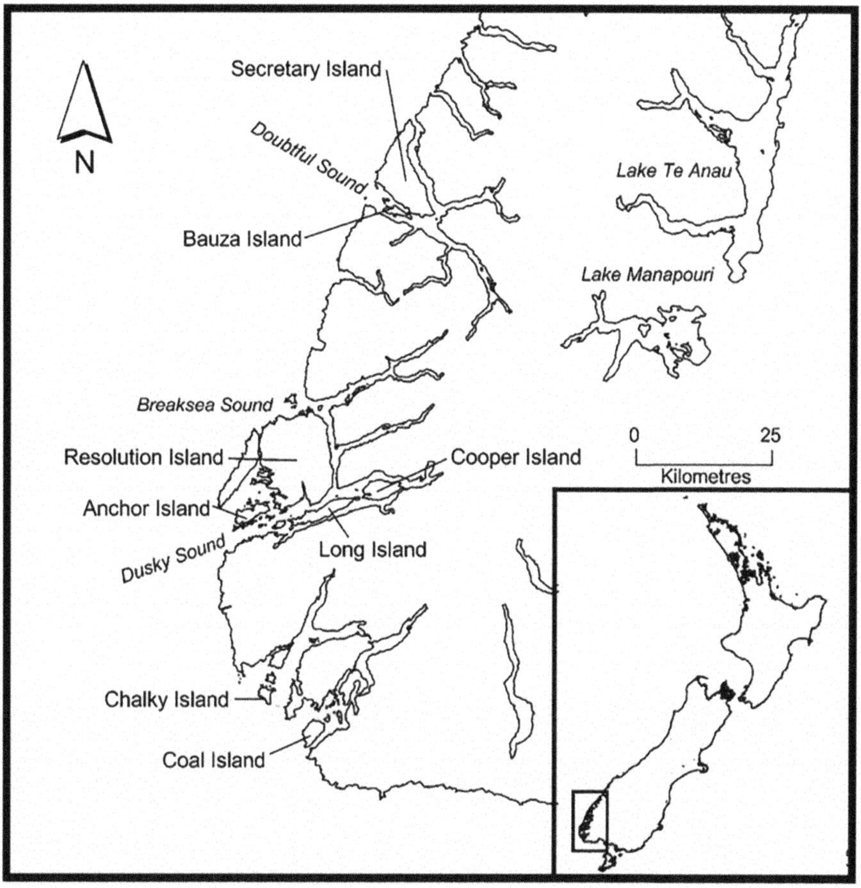

Fig. 7.1 Map of Fiordland islands acting as significant conservation reserves but vulnerable to invasion by stoats. (From Elliott et al., 2010)

eradicated, but 76% were followed by repeat incursions, sometimes more than once. Most of these incursions targeted in-shore islands less than 1500 m from the mainland (e.g., 12 in the Bay of Islands), an easy swim for a stoat (Sect. 7.2), but stoats can certainly swim further than that. Incursions by fertilised females are by far the most dangerous (Sect. 4.5), and the largest islands are the most important and the most expensive to protect.

An early study on Adele Island (87 ha, 800 m offshore) part of Abel Tasman National Park, confirmed that attempts to remove stoats from small islands close to shore has to be an ongoing process. Fenn traps set on Adele in August 1980 caught eight stoats in 50 trapnights within a month, followed by six more over the summer the next year. A year later, another four turned up, but stoat tracks were still observed on the island's beaches. Rowley Taylor and Jocelyn Tilley concluded that trapping can effectively remove naïve resident stoats from islands, but that recolonisation is rapid on any island closer than 1100 m offshore (Taylor & Tilley, 1984). Later data extended this limit to 3–5 km.

Maud/Te Hoiere Island, in the Marlborough Sounds, was one of the first rat-free islands to be prepared as a safe refuge for flightless birds. Unfortunately, the narrow (900 m) crossing to the mainland is a mere hour's paddle for a stoat (Sect. 7.3). Stoats have been removed from Maud three times, in 1983, 1989 and 2003, and they continue to reinvade periodically (Elliott et al., 2010).

Chalky/Te Kakahu In 1999, DOC ranger Murray Willans achieved one of the first successful trap-based, semi-permanent eradications on a large island in Fiordland, Chalky/ Te Kakahu (514 ha) (Elliott et al., 2010). The island itself is 1100 m offshore, but connected to the mainland by a string of smaller islands, all of which were included in DOC's programme (Fig. 7.2). Murray and his team planned a well-timed blitz approach of 140 traps set in June, when nesting sea birds were absent for the winter. The island stoats were completely naïve to traps, and the operation removed 15 residents in the first 2 weeks, plus one more before October. Since then, kākāpō (*Strigops habroptilus*), saddleback/tīeke (*Philesturnus carunculatus*,) yellowhead/ mōhua (*Mohoua ochrocephala*), little spotted kiwi (*Apteryx owenii*), and orange-fronted kākāriki (*Cyanoramphus malherbi*) have been reintroduced to the island. The island remained predator free for the next two decades.

Fig. 7.2 Searching the beaches of Chalky Island/Te Kakahu for stoat footprints or scent cues, during an expedition with a DOC ranger and his stoat dog to check for any survivors after the 1999 eradication, February 2000. (Photo by Author)

The problem is that the isolation protecting the birds is also what makes a predator incursion response so expensive. In August 2022, a single male stoat was found on the island. DOC launched an immediate incursion response, which cost $483, 260 over 8 months, covering management team costs, staff time, conservation dog handlers, helicopter and boat expenses. Then it spent another $210,000 on ensuring the island remained pest-free by installing surveillance systems and doing biosecurity planning.

The report of this incident in the *New Zealand Herald* on 2 April 2024 generated a lot of comment, but, as DOC explained, if the stoat had not been caught it could have killed kākāpō chicks, nationally endangered Te Kākahu skinks (*Oligosoma inconspicuum*), and little spotted kiwi. Professor James Russell told the *Herald* that the expense was justified on the grounds that it is necessary to keep investing in this work, to safeguard the effort and expense already committed.

7.2 Swimming Range of Stoats

There are many other smaller but equally important islands off the Fiordland coast, most covered in untouched forest and well within swimming range of stoats. Islands under 50 ha cannot support a resident stoat population (Veale et al., 2012), so any stoats removed from such small islands could only have been temporary residents or visitors moving between islands as part of their normal wide-ranging foraging strategy. Colonies of burrowing seabirds can survive on small islands because stoats cannot stay after migratory petrels depart at the end of their breeding season (Miskelly et al., 2019). But many questions remained. How could we know how safe an inshore island might be, as refuges from introduced predators in the mainland? More precisely, How far can stoats swim?

Stoats are related to otters and mink, which are certainly good swimmers, so the question of how far stoats can swim is an important consideration in choosing which islands are far enough offshore to be safe from invasion by swimming stoats. For a long time the safe distance was taken to be about 1.5 km, merely from the distribution of the islands on which resident stoats have been seen. But that didn't tell us how often stoats could make shorter visits to closer islands too small to support a permanent population or how far off an island has to be to be safe from future incursions.

Captive Observations The Aquatic Research Centre at Waikato University has a flume, a glass-sided channel with a motor that can run an endless current round a circuit at a given speed. We bought ten wild-caught stoats from Landcare Research at Lincoln, where they had been used for a different experiment and were well habituated to captivity. One of them was obviously sick, so was rescued from the water within minutes. This animal had an advanced pulmonary disease, which is common in New Zealand (McDonald et al., 2008). The other nine were filmed through the glass sides of the flume, and from underneath so as to watch their paddling action (Fig. 7.3).

Fig. 7.3 This female stoat swam at a steady 1 km/h for nearly 2 h against a constant current in a flume, January 2013. (Photo by Author)

Four of five males and two of four females could hold a position against a constant current of 1.35 km/h for at least 5 min. In steady swimming against a current of 1 km/h, they all paddled with all four paws at around 250 strokes per minute, most strongly and steadily spreading the front paws like an otter. Four of the nine swam nonstop for more than an hour, including one female who swam for nearly 2 h, covering 1.8 km (King et al., 2014).

Stoats have turned up on Kapiti Island (Prada et al., 2014) and Rangitoto (Veale et al., 2012), both well outside previous estimates of their swimming range. Could they have got there unaided? If assisted by a current or a floating log, it would perhaps not be impossible for wild stoats to swim further than those captive ones did. The implication is that many islands, especially the closer ones up to 3–5 km offshore, are at permanent risk of periodic visits or re-invasions by stoats. DOC has since updated its register of islands suitable to host translocations of vulnerable native species.

7.3 Resolution/Tau Moana and Secretary/Ka Tū-waewae-o-Tū Islands

Trap-based campaigns can work well on smaller islands, but not on larger ones (>1200 ha), on which the chances of a stoat never encountering a trap, or refusing to enter it, are higher. Resolution and Secretary are the largest and most important rat-free islands off the Fiordland coast (Fig. 7.1). They have disproportionate value for conservation, and they offer the chance to distinguish the effects of stoat predation from that of rats. But both are well within the swimming range of stoats, and both are remote and rugged areas where field work is extremely difficult. Both support valuable protected biota, thereby complicating decisions about the use of toxins.

Stoats were first recorded on Resolution Island (20,800 ha) by Richard Henry (Hill & Hill, 1987). Their initial impact can only be imagined by us, but Henry saw it in real life. He first saw a stoat (he called it a weasel) in 1900, then describes not finding any saddleback or piopio in 1901 (Henry et al., 2022: 245). His descriptions of the pitiful silence of the forest after the mustelids arrived is powerful evidence of an unmanaged catastrophe, and not only on the island. Depressed, he handed in his resignation in 1902, but quickly withdrew it. But by 1904 he was saying that he hadn't seen a saddleback or piopio in years.

No attempt to manage stoats on Resolution was made until a trapping network was first set there in July 2008, eventually numbering 2352 traps at c. 200 m intervals. From then until January 2015, 644 stoats were caught, and others still survive after 6 years and 24 trap sessions. Detailed modelling (Anderson et al., 2016) suggested that the 419 collected in the first catch represented an undisturbed overwinter population of 449 stoats before the first recruitment cycle. A simulation of the trapping regime estimated a 0.21 probability of eradicating stoats from the island by 2019; if it ceased, the original numbers could be restored within three breeding seasons.

The model implied that the high fecundity of adult females, their resistance to entering traps, and the arrival of occasional immigrants swimming across the narrow Acheron Passage, decrease the chances of success unless DOC's operational budget is at least doubled. The funds available for Resolution in the 2021/22 fiscal year totalled $16,0672, with no provision for expansion or systematic outcome monitoring.

On Secretary Island (8140 ha) an intensive effort starting with 945 trap sites also could not eradicate stoats. In July 2005, 95 stoats were caught, plus 9 caught in November that would have been there in July. Another 44 were killed in February 2006, and up to 9 every trapping session since then (McMurtrie et al., 2011).

But even without a complete eradication, trapping has suppressed stoat numbers on both islands down far enough to benefit threatened native species, and allow some reintroductions of native species. On Secretary Island, northern tokoeka (*Apteryx australis*), kākā, weka (*Gallirallus australis*), forest birds, invertebrates, general bird counts and Fiordland skinks have all increased, and rock wren/pīwauwau (*Xenicus gilviventris*)/and mōhua have been released. On Resolution Island, Fiordland skinks, pīwauwau and mātātā/fernbird (*Poodytes punctatus*) have increased, a forest Gecko has been found; mōhua and SI saddleback/tieke have been successfully introduced.

Unfortunately, there have also been recent incursions of rats on both islands, an alarming outcome attributed by some to the reduced number of stoats. Is it possible that stoats can keep rats off these precious island refuges? If true, this seems to confirm that, in nature, managers cannot change only one thing.

7.4 Waiheke Island

Like Resolution and Secretary, Waiheke is a large island (9324 ha) on which an intensive and well documented trapping effort has not, so far, succeeded. But unlike the remote and uninhabited Resolution and Secretary Islands, Waiheke is effectively an offshore suburb of Auckland (21 km away by ferry). It is also well outside the swimming range of stoats except via nearby Ponui Island, 5.4 km from Auckland. The ancestors of the current population of stoats might have got to Waiheke from Ponui, but there are none there at the moment, and no genetic evidence that Waiheke stoats have been supplemented from Ponui in recent times (Veale, 2024). So if eradication could ever be achieved, the risk of reinvasion could not be eliminated, but would be minimal.

The problem is that the last few surviving resident stoats are extremely resistant to eradication by trapping, and toxins are not acceptable. Waiheke's experience confirms that stoats vary in personality—the bold ones are the most quickly removed (Johnstone et al., 2023). The older, more cautious ones are innately unlikely to engage with any human devices until they have higher priorities—like in the breeding season. It certainly helps to design a new eradication attempt to remove them all while they are still naïve—as on Chalky/Te Kakahu in 1999—but there will always be a few left that require trappers to settle in for a lot of experimentation over long campaign, or else try another approach.

For example, a network of more than 600 traps on pest-free Motutapu Island (1509 ha) in the Hauraki Gulf, aimed to catch an incursion of stoats first detected in May 2020. In September 2020 and January 2021, two were caught on Motutapu or the adjacent island Rangitoto, but a third evaded traps for nearly a year. Finally, DOC brought in a new lure, a fake den equipped with a playback recording of baby stoats squeaking, and female stoat-scented bedding with carcases. Males searching for mates can seldom resist such a cue (Sect. 4.5), so the last stoat on Motutapu was caught during the peak of the breeding season in November 2021 (https://www.doc.govt.nz/news/media-releases/2021).

The resident human population of Waiheke is very interested in conservation and in supporting and the aim of eradicating stoats. First started in 2016, the campaign now involves more than 40 neighbourhood predator control groups operating more than 1600 traps. Detection cameras and predator dogs, and a network of sharp-eyed human observers, are quick to report and search the locations of any surviving stoats.

From February 2020 to January 2025, a total of 230 stoats were removed from Waiheke Island (Holmes, 2025), and every one has been sent for genotyping. The results show that the resident stoats are all seriously inbred, confirming the complete lack of any immigrants from outside, plus details of the family relationships of each captured individual (Veale, 2024).

Plotting the unpublished genetic data against capture locations shows the distributions of each family group, and how its members are well aware of each other. Some dispersing young have been shown to travel long distances (at least 7.3 km) through the dense trapping network, completely ignoring all the many trap stations they pass (Fig. 7.4).

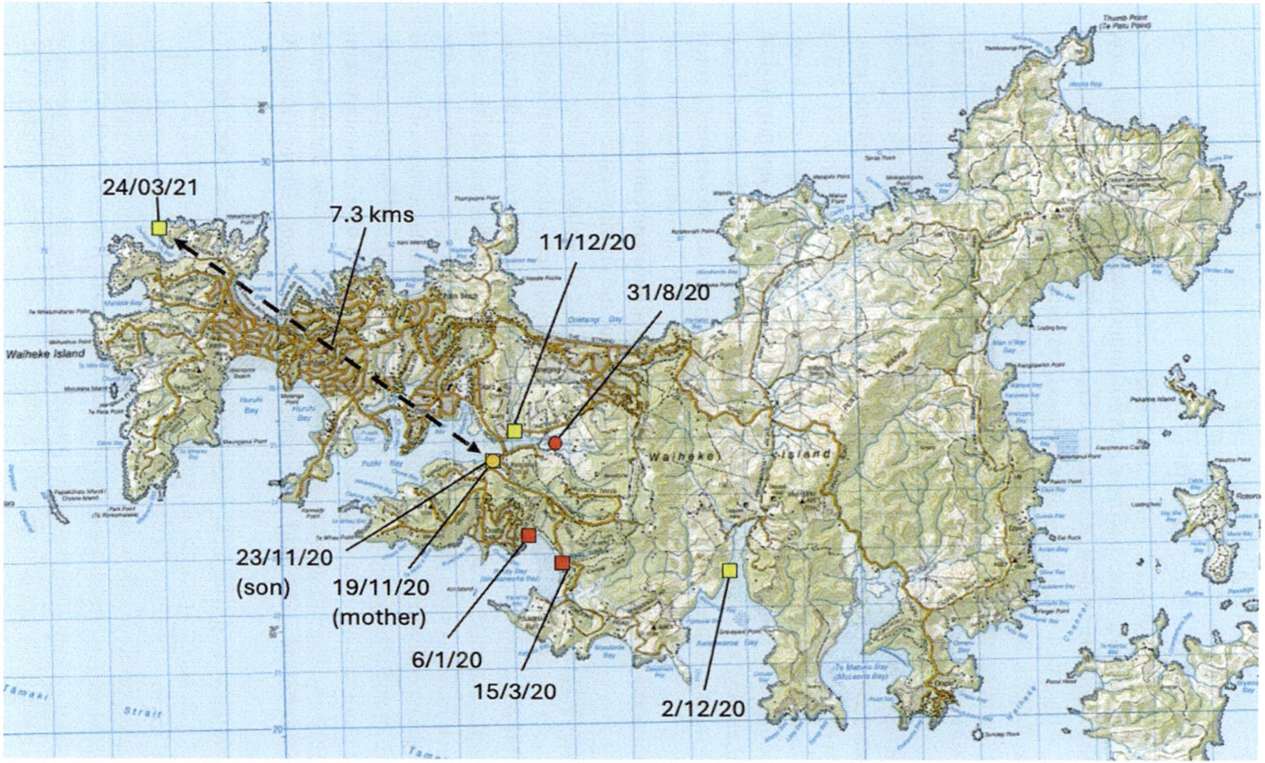

Fig. 7.4 The distribution of two litters of stoats born on Waiheke Island during the current eradication program. Dates of capture for each stoat are recorded, males as squares, females circles. The first litter (in red) was born in September 2019 and all known litter mates were caught within 2.5 km of each other. A daughter from that litter (coloured orange) was caught after having her own litter (coloured yellow). Her smell remained in the trap where she was caught, and one of her sons was then caught in the same trap 4 days later. Her other offspring then dispersed across the island; one of them was caught 4 months later, 7.3 km away (through a dense network of traps) from where the mother was caught. (Map and data from Andrew Veale)

One large male stoat was found dead on 1st September 2023 on the grass near the Awaroa wetland. There was no obvious cause of death, or any injuries, and although testing found traces of anticoagulants in his liver, these were at levels unlikely to have caused death. He was found to be almost 4 years old, having avoided traps his whole life. He had fathered 75% of the known litters born in 2023, and had also fathered other litters in previous years. This stoat has long maintained the Waiheke stoat population, but was never caught (Veale, 2024).

Another male stoat was detected living in a woodpile near the Rangihoua wetlands by public sightings and detector dogs. This woodpile was surrounded by a range of traps with a variety of baits, and eventually he was caught in a live trap. He was then moved to the Manaaki Whenua Landcare Research animal facility where he was tested for behavioural traits. Of all male stoats tested, he was the most neophobic and cautious. Within any wild population there will be bold individuals and more cautious ones; control programs rapidly favour those that are more timid and cautious, either by inheritance or learned aversion.

The current debate concerns how long the local organiser tekorowaiowaiheke.org should continue hoping that trapping alone might finish the job, or if not, when and how to decide whether the use of toxins might become practicable. If never, what might be the alternative? Meanwhile, there is little doubt that suppression of stoats on Waiheke by traditional means is bringing real benefits for the island population of kākā, and increases in average annual counts of native and endemic birds (Sect. 14.2.7).

7.5 Orkney Islands

The Orkneys are a rugged group of islands off the north coast of Scotland, inhabited by 22,500 people and unknown numbers of stoats. Stoats are native to Britain, but not to Orkney, which has been hitherto free of predatory land mammals. Stoats are treated as invasive pests threatening the islands' important seabird colonies.

Stoats were introduced to the Orkney archipelago around 2010. In 2016, a project aiming to eradicate them to preserve ecologically and economically important native wildlife was designed and implemented. Expertise from New Zealand was imported to help protect the Orkney Native Wildlife (https://www.orkneynativewildlife.org.uk).

The programme started with a "knockdown" phase, followed by a "mopping-up" phase to remove stoats that escaped capture. It has distributed hundreds of DOC 150 and DOC 200 traps, which have killed some 5000 stoats so far. It uses proprietary scent lures made by Manaaki Whenua Landcare New Zealand, and deploys Mostela camera boxes (Mos & Hofmeester, 2020).

Zub et al. (2022) used data from this project to iteratively predict the progress toward eradication. They applied spatially explicit individual-based models to estimate the proportion of stoats being exposed to capture, and then compared these simulation-based predictions with removal data, to estimate changes in the population size through time. They also used sighting data from members of the public to refine the probability of eradication.

As the programme progressed, the information derived from different types of data and quantitative models suggests that an initial knockdown phase was achieved at least in some areas. There was spatial and temporal heterogeneity in the distribution of captures, probably due to stoats moving about in response to disturbance, and to trap shy stoats being exposed to capture but not caught. This heterogeneity calls for additional data sources (e.g., from camera traps or detection dogs) to estimate the proportion of trap-shy individuals and the size of the untrapped population. The model tracked the effectiveness of current trapping approaches to help to inform project managers when the project achieved the knockdown phase milestone, and ultimately the feasibility of eradication. It will be interesting to see how long it takes, compared with the ongoing effort to eradicate mink from Scotland.

References

Anderson, D. P., McMurtrie, P., Edge, K. A., Baxter, P. W. J., & Byrom, A. E. (2016). Inferential and forward projection modeling to evaluate options for controlling invasive mammals on islands. *Ecological Applications, 26*(8), 2546–2557. https://doi.org/10.1002/eap.1415

Clout, M. N., & Russell, J. C. (2006). The eradication of mammals from New Zealand islands. In F. Koike, M. N. Clout, M. Kawamichi, M. De Poorter, & K. Iwatsuki (Eds.), *Assessment and control of biological invasion risks* (pp. 127–141). IUCN.

Corson, P. (2024). Personal Communication.

Elliott, G., Willans, M., Edmonds, H., & Crouchley, D. (2010). Stoat invasion, eradication and re-invasion of islands in Fiordland. *New Zealand Journal of Zoology, 37*(1), 1–12.

Henry, R. T., Hill, S., Hill, J., & Jaenecke, V. (2022). *Letters of a naturalist. The field accounts of Richard Henry on Resolution Island.* Putangi Publications.

Hill, S., & Hill, J. (1987). *Richard Henry of Resolution Island.* John McIndoe.

Holdaway, R. N., Worthy, T. H., & Tennyson, A. J. D. (2001). A working list of breeding bird species of the New Zealand region at first human contact [Review]. *New Zealand Journal of Zoology, 28*(2), 119–187. WOS:000170520800001.

Holmes, J. (2025). tekorowaiowaiheke.org.

Johnstone, K. C., Garvey, P., & Hickling, G. J. (2023). Invasive mammal control selects for trap-recalcitrant behaviour and personality. *Biological Invasions.* https://link.springer.com/article/10.1007/s10530-023-03191-4

King, C. M., Veale, A., Patty, B., & Hayward, L. (2014). Swimming capabilities of stoats and the threat to inshore sanctuaries. *Biological Invasions, 16*, 987–995. https://doi.org/10.1007/s10530-013-0564-2

McDonald, R. A., Birtles, R. J., McCracken, C., & Day, M. J. (2008). Histological and serological evidence of disease among invasive, non-native stoats *Mustela erminea. Veterinary Journal, 175*(3), 403–408. https://doi.org/10.1016/j.tvjl.2007.01.028

McMurtrie, P., Edge, K. A., Crouchley, D., Gleeson, D., Willans, M. J., & Veale, A. J. (2011). Eradication of stoats (*Mustela erminea*) from Secretary Island, New Zealand. In C. R. Veitch, M. N. Clout, & D. R. Towns (Eds.), *Island invasives: Eradication and management* (pp. 455–460). IUCN.

Miskelly, C. M., & Fraser, J. R. (2006). Campbell Island snipe (*Coenocorypha* undescribed sp.) recolonise subantarctic Campbell Island following rat eradication. *Notornis, 53*, 353.

Miskelly, C. M., Bishop, C. R., Taylor, G. A., & Tennyson, A. J. D. (2019). Breeding petrels of Chalky and Preservation Inlets, southern Fiordland-a test of the 'refugia from resident stoats' hypothesis. *Notornis, 66*, 74–90.

Mos, J., & Hofmeester, T. R. (2020). The Mostela: An adjusted camera trapping device as a promising non-invasive tool to study and monitor small mustelids. *Mammal Research*. https://link.springer.com/article/10.1007/s13364-020-00513-y

Prada, D., et al. (2014). Unwelcome visitors: Employing forensic methodologies to inform the stoat (Mustela erminea) incursion response plan on Kapiti Island. *New Zealand Journal of Zoology, 41*, 1–9.

Russell, J. C., & Clout, M. N. (2004). Modelling the distribution and interaction of introduced rodents on New Zealand offshore islands. *Global Ecology and Biogeography, 13*, 497–507.

Taylor, R. H., & Tilley, J. A. V. (1984). Stoats (*Mustela erminea*) on Adele and Fisherman Islands, Abel Tasman National Park, and other offshore islands in New Zealand. *New Zealand Journal of Ecology, 7*, 139–145.

Towns, D. R., Bellingham, P. J., Mulder, C. P. H., & Lyver, P. O. B. (2012). A research strategy for biodiversity conservation on New Zealand's offshore islands. *New Zealand Journal of Ecology, 36*, 1–20.

Veale, A. J., et al. (2012). Genetic population assignment reveals a long-distance incursion to an island by a stoat (*Mustela erminea*). *Biological Invasions, 14*(3), 735–742.

Veale, A. J. (2024). Personal Communication.

Veale, A. J., Hannaford, O. D., Russell, J. C., & Clout, M. N. (2012). Modelling the distribution of stoats on New Zealand offshore islands. *New Zealand Journal of Ecology, 36*(1), 38–47.

Zub, K., Garcia-Diaz, P., Sankey, S., Eisler, S., & Lambin, X. (2022). Using a modeling approach to inform progress towards stoat eradication from the Orkney Islands. *Frontiers in Conservation Science*. https://doi.org/10.3389/fcosc.2021.780102

Toxins

8

Abstract

Registrations for new vertebrate pesticides have declined over the last 40 tears in most parts of the world, except in New Zealand. The ongoing need to protect our native biodiversity against invasive predators makes us one of the few countries where landscape-scale routine use of toxins in conservation areas is officially accepted. The most advanced and strictly monitored protocols have been developed over decades of experience, especially for aerial distribution of 1080 (sodium monofluroacetate). Public criticism has driven research demonstrating that a well-timed 1080 drop can clear large areas of rats and stoats by direct and secondary poisoning much more effectively than can trapping. Before- and after-monitoring of radio-collared stoats in Pureora Forest Park during a 1080 drop in 1997 proved that all 13 collared stoats died, most having eaten dead rats. Successive landscape-scale programmes distributing 1080 in Fiordland in seedfall years increased the survival of sitting females and newly fledged juveniles of hole-nesting bush birds. The resulting relief from nest predation is temporary, but adds valuable cohorts of young adults to the breeding populations. New alternative toxins like PAPP aimed at stoats and cats are in development.

Keywords

Sodium monofluroacetate (1080) · Secondary poisoning · Radiocollared stoats · Nest predators · PAPP

Deliberate use of toxins on sentient animals is unpleasant to think about, but the looming extinction of New Zealand's irreplaceable endemic fauna is arguably

C. M. King, *Stoat in the Dock*, Fascinating Life Sciences,
https://doi.org/10.1007/978-981-96-4562-6_8

worse. The reasoning used to justify it is that the costs in ecological and animal welfare of using toxins against pests are much less than the costs in damage to protected resources if toxins are not used. That idea does not apply in other countries, where the risk to valuable native mammals is far too great. In New Zealand, the only native land mammals are two species of insectivorous bats, both well adapted to hunting on the ground, but no longer common there. The only other mammals likely to pick up a toxic bait are all introduced pests. Consequently, the side effects of poisoning operations are less a cause for concern to managers here than it is elsewhere, and in New Zealand toxins have remained permissible tools for stoat control long after most other countries have banned them.

The 1080 Debate Of the several toxins registered for pest control in New Zealand, the most frequently used is 1080 (sodium monofluroacetate). 1080 is a natural plant toxin widespread in native vegetation in Western Australia, and breaks down quickly in soil and natural water. Most mammals (except those endemic to western and central Australia) are highly sensitive to 1080, so it can be distributed in tiny doses hidden in small baits made deliberately uninteresting to birds.

1080 is not only remarkably efficient in reducing the numbers of possums and rats: it is also much more efficient than are traps in removing stoats. Stoats are particularly susceptible to minute amounts of 1080 (1 mg in a 1 g bait is a fatal dose for stoats of all sizes). They seem not to be able to detect it, and a lethal dose acts quickly (on average within 3 h, depending on the method of delivery) and apparently humanely (Potter et al., 2006). Ferrets are larger animals and can recover from a dose of 1 mg/kg body weight, but more than that is fatal (Marshall, 1963).

So, given certain well-understood and fiercely monitored precautions, aerial distribution of 1080 over huge areas of New Zealand is by far the most effective means of stoat control at landscape scale. The baits used are not meat-based, so are not themselves taken by stoats, but a moribund rat or possum staggering about carrying a lethal dose of 1080 is defenceless against a stoat. In turn, stoats find such an easy prey completely irresistible, and they do not reject fresh carcases either. That makes stoats very vulnerable to secondary poisoning. Since 1964, huge, co-ordinated operations funded by agricultural authorities have cleared most forests of tuberculous possums, and no doubt killed uncounted numbers of stoats incidentally.

In August 1997, confirmation of this guess came from a large area of forest (8577 ha) at Waimanoa, in Pureora Forest Park. Before a scheduled aerial 1080 operation, 13 resident stoats were fitted with radio-collars (Figs. 8.1, 8.2). Every one of them died. Others not examined were probably also affected, because no mustelids were tracked or trapped in that area for the next 3 months (Murphy et al., 1999).

Years of repeated and reliable research show that carefully managed aerial distribution of 1080 is the most efficient weapon for landscape-scale pest control, because a single aerial operation can achieve complete and drastic knockdowns of the three most significant predators of forest birds, at least over the breeding season. In forested country, well-timed 1080 operations bring temporary benefits for wildlife

Fig. 8.1 A live, unconscious stoat fitted with a radio collar and ear tag by Craig Gillies. (Photo by Nic Gorman, DOC Image Library)

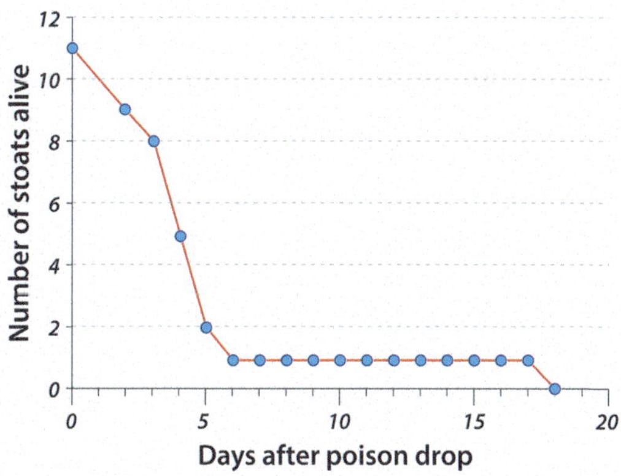

Fig. 8.2 Over the next 2–18 days, all 13 of the radio collared stoats were accounted for. The daily survival of the eleven that died within 18 days was recorded, and another two were found dead. One had an empty gut, but of the other twelve, eight had recently eaten a rat, and another one had eaten a dead possum. All but one had residues of 1080 in the tissues. The area remained free of mustelids for another 3 months. Hence, a single 1080 drop can simultaneously remove the three most significant predators of forest birds. (Redrawn from data supplied by Murphy et al., 1999)

comparable with mass reproductive failure of rats and stoats, at least over limited areas and at repeated and great costs. One-off operations deliver no long-term increase in density of protected species, because the rats and stoats can come back within 6 months.

Monitoring the results of a 1080 operation provides proof of short-term but critically important relief for threatened populations of birds (Kilner et al., 2024). The acknowledged loss of some individuals is vastly outweighed by the boosted productivity of the breeding stock. 1080 drops protected the sitting females of hole-nesting birds during the Battle for the Birds programme and their chicks, rejuvenating the adult breeding population (Sect. 14.2.3) and reduced the mortality of adult kiwi at Tongariro Forest Kiwi Sanctuary (Sect. 14.2.1). Local accumulations of otherwise irremovable, untrappable stoats can be eliminated with a short burst of 1080 in bird-proof tunnels (Robertson et al., 2016).

Objections to 1080 There are two problems. One is that these benefits don't last long, because rats come back quickly, and in greater numbers than possums, and can stay higher than usual until possums recover. The other is that 1080 is also fatal to mammals that other people value, especially dogs and deer (pigs are less affected because they can regurgitate it, but there is still a risk of tainted meat).

Hence, the use of 1080 is subject to ongoing and passionate protests by well-organised lobby groups aiming to have it banned (Hansford, 2016). The argument is reviewed by New Zealand's statutory Parliamentary Commissioner for the Environment (PCE), who is charged with researching and offering independent advice on matters of environmental concern. PCE reports are regarded as authoritative and unbiased analyses of many knotty problems in that often controversial field. So the use of 1080 was an obvious topic for examination by the PCE.

In 2011, Jan Wright, then the current PCE, published her analysis of this hugely difficult question (Wright, 2011). The arguments against use of 1080 have gathered a lot of public support, despite many careful scientific studies and multiple critical reviews all concluding that the objections to the use of 1080 do not outweigh the benefits. But, far from supporting calls to ban 1080, Wright suggests that we should use more of it. This analysis, certainly the most authoritative and thorough to date, did not end the debate. Why not?

Mainly because, although the science of 1080 use is settled, the objections are more personal. Managers have to find ways of acknowledging different viewpoints while getting on with the business of efficient protection of native species. A revealing parallel case concerning Māori objections to the eradication of kiore (*Rattus exulans*, a tribal treasure to some groups) from Little Barrier Island/Hauturu is sympathetically described by David Towns (Towns, 2023: 257–263).

Questions surrounding the legitimate use of 1080 continue to rank high among the most intractable dilemmas facing managers of large forests in New Zealand. No-one regards 1080 as ideal—indeed, all parties, including its current supporters, hope that better methods will be found eventually—but for now, 1080 is the most effective option available (Eason et al., 2017). The problem is urgent, because eradication of stoats on the mainland would be impossible without toxins (Veale, 2024).

Alternative Toxins Para-aminopropiophenone (PAPP) is a new toxin being developed in New Zealand specifically for the control of stoats and feral cats. It takes advantage of the special sensitivity of these species to PAPP, but research on new agents is expensive. PAPP paste was approved as a stoat control agent in New Zealand in August 2011, and meat baits containing toxic doses of PAPP distributed in bait stations in the field have reduced stoat numbers rapidly (Dilks et al., 2011). Wider uptake has been slowed by label restrictions, as the paste needs to be inserted into freshly hand-made meatballs and is difficult and messy to use (Murphy, 2025). PAPP is more humane than 1080, but on the other hand, a tool aimed specifically at carnivores does not achieve the welcome secondary effects of 1080 in removing rats at the same time.

Two new meat stoat baits are being delivered that could be aerially applied, one containing 1080 and the other PAPP. Hand-laid trials of these baits have been successful, and aerial trials are planned for registration purposes (Rickett et al., 2023). Preliminary field trials have also been undertaken investigating re-setting toxin devices for stoats with PAPP and 1080 (Murphy, 2025).

There is still a lot of argument about the use of toxins against mammalian pests in New Zealand. The dilemma is that in places where clearing out stoats is most urgently needed, trapping isn't quite enough but toxins are ruled out by conflicts with human interests. On Waiheke Island (Sect. 7.4), long term trapping suppresses stoat numbers well enough to allow the recovery of huge flocks of North Island kākā (*Nestor meridionalis septentrionalis*), but the last few stoats are very reluctant to engage with any human devices. On Chalky/Te Kakahu Island (Sect. 7.1) the last stoat remaining after the second (1999) eradication had to be shot. Ongoing research connected with the Predator Free New Zealand programme (Chap. 13) might be able to find an answer in future.

References

Dilks, P., Shapiro, L., Greene, T., Kavermann, M. J., Eason, C. T., & Murphy, E. C. (2011). Field evaluation of para-aminopropiophenone (PAPP) for controlling stoats (*Mustela erminea*) in New Zealand. *New Zealand Journal of Zoology, 38*, 1–8.

Eason, C. T., Shapiro, L., Ogilvie, S., King, C., & Clout, M. (2017). Trends in the development of mammalian pest control technology in New Zealand. *New Zealand Journal of Zoology, 44*, 267–304.

Hansford, D. (2016). *Protecting paradise: 1080 and the fight to save New Zealand's wildlife*. Potton & Burton.

Kilner, C. J., Hunter, C. M., Cole, R. E., Allan, T. E., Rawlence, T. E., Tinnemans, J. S. J., Bell, M. A. N., Bell, C. W., Malham, J. P., McDonald, A., Moran, K. J., & Elliott, G. P. (2024). Rifleman (*Acanthisitta chloris* sp.) population responses to aerial 1080 (sodium fluoroacetate) predator control in beech forests. *New Zealand Journal of Ecology, 48*(1), 3577.

Marshall, W. H. (1963). *The ecology of mustelids in New Zealand* (DSIR information series 38). DSIR Publishing.

Murphy, E. C. (2025). Personal communication.

Murphy, E. C., Robbins, L., Young, J. B., & Dowding, J. E. (1999). Secondary poisoning of stoats after an aerial 1080 poison operation in Pureora Forest, New Zealand. *New Zealand Journal of Ecology, 23*, 175–182.

Potter, M. A., Barrett, D. P., & King, C. M. (2006). Acceptance by stoats (*Mustela erminea*) of 1080 (sodium monofluoroacetate) in small-volume baits and its effect on behaviour and time to death. *New Zealand Veterinary Journal, 54*, 350–356.

Rickett, J., Wallace, P., Murphy, E., & Galloway, T. N. H. (2023). Trials with non-toxic baits for stoats and feral cats. *New Zealand Journal of Ecology, 47*(1), 3505.

Robertson, H. A., Craig, E., Gardiner, C., & Graham, P. J. (2016). Short pulse of 1080 improves the survival of brown kiwi chicks in an area subjected to long-term stoat trapping. *New Zealand Journal of Zoology, 43*, 351–362.

Towns, D. (2023). *Ahuahu. A conservation journey in Aotearoa New Zealand.* Canterbury University Press.

Veale, A. J. (2024). Personal communication.

Wright, J. (2011). *Evaluating the use of 1080: Predators, poisons and silent forests*. Parliamentary Commission for the Environment.

Remote Monitoring

9

Abstract

Remote monitoring uses footprints, radio telemetry and trail cameras to follow the activities of wild animals without detaining them. Footprint tracking tunnels provide a simple and cheap method of monitoring the presence/absence of small mammals. The field protocols have been refined and standardised over decades of experience into an informative tool for use at landscape scale. Radio telemetry for small mustelids is more difficult, because of their tube-shaped bodies, but radio tracking data are especially revealing. Telemetry data are not confined to fixed locations (as livetraps, tracking tunnels or cameras are) so can follow marked animals to unexpected places, so long as they stay in range. Radio locations can define home ranges, which is useful information when planning control of ground pests. Equipping stoats and ferrets with radios before a control operation can define the number removed by professional trappers or by secondary poisoning. Remote cameras can identify nest predators, incursions onto islands, and breaches in pest-exclusion fences, but small, fast-moving mustelids may still be missed unless detained by an attractive lure. An advanced smart bait dispenser with accompanying detector confirmed that radio-collared ferrets (and probably stoats) need time to engage with human devices.

Keywords

Tracking tunnels · Radio telemetry · Trail cameras · Fenced sanctuaries · Scentinel

The use of remote monitoring to follow the activities of wild animals without detaining them has had a long history in New Zealand.

C. M. King, *Stoat in the Dock*, Fascinating Life Sciences,
https://doi.org/10.1007/978-981-96-4562-6_9

9.1 Footprint Tracking

One spin-off from my failure to observe live stoats in the Orongorongo Valley (Sect. 3.1) was the germ of an idea for an alternative system of monitoring stoats in rugged country unsuitable for live-trapping. In cooler countries and seasons, snow provides a ready-made tracking medium in which to record footprints. For the rest of us, it might be possible to provide one.

My first idea was to adapt the two-component dye technique used by police to take finely detailed fingerprints. It developed into an independent design for cheap, home-made tunnels and ink pads for field use (King & Edgar, 1977; King et al., 2015). The system worked and was affordable for unfunded work, but clunky. Fortunately, the idea was taken up by entrepreneur Warren Agnew, and developed into a commercially successful system that has been widely used throughout the country (Fig. 9.1). Agnew's company, Gotcha Traps NZ, supplies simple "Black Trakka" tunnels and pre-inked papers, along with a guide to identifying animal tracks and a lot of background information https://gotchatraps.co.nz.

Artificial tunnels containing tracking papers and ink pads, set out in grids or lines, have two main uses. First, tracking tunnels have always been especially good for simple presence/absence surveys in remote places, because they offered the first simple and easy way to answer one of the most important and difficult questions about any pest species: Where are they? Or conversely, are there any here? Students planning to begin a field study need to know where their target animals are and, if possible, roughly how many of them are there.

Second, routine population monitoring has for a long time been done more easily and cheaply by tracking tunnels than by traps. Provided certain strict conditions are met, a rough index of abundance can be calculated from the proportion of tunnels recording stoat tracks over a given short period (Brown et al., 1996). Baited tracking

Fig. 9.1 Black Trakka tunnels have been deployed across conservation areas for decades. The footprints of stoats, rats and mice, lizards and insects are easily distinguishable. (Photo by Marcus Agnew)

tunnels laid out by compass across randomised transects and operated under a fixed, standardised protocol developed by the Department of Conservation (DOC) and monitored by Craig Gillies (Gillies & Williams, 2013) have been used since the 1990s to check the distribution of pests and their reduction in numbers after a control operation (Murphy et al., 1998).

These indices are broadly helpful, but they have limitations. First, if more than one track appears on the paper, there is no way to tell whether they were made by one animal visiting several times or more than one visiting once each. Second, the results are less accurate if the papers are not changed frequently. The standard protocol originally required tunnels to be withdrawn after 3 days, but that resulted in databases with too many zeroes. Since 2015 the tunnels have been running for 21 days (Walker, 2025).

Advantages of Tunnel Tracking The great value of footprint tracking is that the equipment is cheap and easy to operate. More important, the sheer number of standardised surveys reported over decades (and ongoing) is proving a remarkable resource for retrospective research enabled by new developments in advanced statistics (Chap. 11).

Recently, more comprehensive stats packages have enabled all sorts of meta analyses yielding ever more astonishing results from long-term tracking tunnel data. Walker et al. (2019) fitted models to 18 years of 3-monthly tracking records from a largescale tunnel network laid out under the Department of Conservation's randomised protocol to (1) describe and classify forest-wide spatial patterns of temporal dynamics of unmanaged rat and mouse populations, and (2) understand the population-limiting roles of environment and biotic interactions. The team distinguished six classes of forest rodent dynamics grading from 'irruptive', with low median but synchronous high maximum rat and mouse tracking rates in colder forests, to 'continuously ratty', with high median rat and low unsynchronised mouse tracking rates in warmer forests. Part III of this book describes two pioneering field studies chosen, by chance, to represent the opposite ends of this spectrum.

The huge spatial and temporal span of tracking tunnel data available to Walker's team suggested that, in general, ship rats and house mice have an unstable balance of advantages. Ship rats are more sensitive to lower temperatures than are mice. They survive less well in cooler habitats, as would be expected from a species that evolved in tropical climates. Craigieburn Forest Park was the coolest and highest-elevation sampling site included in the National Parks Survey (Chap. 4) and ship rats were absent there until the warming climate enabled their invasion in about 2010 (Harris et al., 2022).

By contrast, house mice are much more tolerant of cool climates, or else small enough to be able to find shelter in small spaces, so are widely distributed down to the subantarctic islands. On the other hand, ship rats hunt and kill mice whenever possible, so the activities of mice are dominated by the need to avoid ship rats, and that limits their foraging time wherever rats are abundant (Bridgman et al., 2013, 2018). Later analyses of the same tracking tunnel datasets produced yet more detailed insights at geographic scale (Chap. 11).

I never expected my invention of footprint tracking tunnels, my parting gift to DSIR in 1977, to produce so much information over such a long term. It was a good thing I was not allowed to patent it (public servants were not then allowed to make personal profits from state-funded research), which would have restricted use of it by underfunded programmes. I am astonished to see how the idea has been spread and improved since my early prototype, and what modellers have made of the mountains of simple data it produced.

New tools developed more recently, like chew cards and Goodnature toothmark cards (https://goodnature.co.nz/products/trap-location-cards) can do the same job as tracking tunnels. Both can be bought cheaply from manufacturers and laid out in larger quantities than tunnels, according to a strategic plan. Animal visitors are identified by their distinctive tooth marks (Sweetapple & Nugent, 2011).

Remote cameras (Sect. 9.4) are better still, and far more informative. But they are also much more expensive, demand a lot of time and expertise to download and process/interpret the images, and are attractive to thieves. In the future, arrays of footprint tracking tunnels or chew cards might give way to remote cameras, for projects where vastly more data is required and can be processed with AI tools. For simpler questions or unfunded surveys, tracking tunnels are still useful, at least for the moment. In this and other fields depending on volunteer labour and scarce finances, cheap and simple technology has a long term advantage.

9.2 Radio Telemetry 1: Weasels in Europe

Live trapping is a crude and disruptive way of following the activities of live animals, because it inevitably distorts the animals' natural movements, but capture-mark-release was for years the only method available (Sect. 2.1). It was not until the 1990s that the development of radio telemetry allowed researchers a way to limit dependence on traps to the necessary initial capture, plus occasional battery changes.

Helpful Information from Europe Weasels are usually not abundant enough in New Zealand to study by radiotracking, but seem to be increasing in places where stoats have been well suppressed (Sect. 10.3). Weasels are not just miniature stoats; they differ in body size and home range extent, in diet and dispersal range, and—most importantly—in reproductive capacities. So there may come a time when we might need some clues from European studies about the home range size, travel distances, daily activity and population dynamics that could be applicable to New Zealand weasels too.

Weasels and stoats have slim, tube-shaped bodies, so it is very difficult to attach a radiocollar firmly enough to stay on without strangling them. The first people to try were Wlodek Jędrzejewski and Bogamilla Jędrzejwska, working in Bialowieza National Park, eastern Poland. They followed twelve radio-collared weasels over the 2 years 1990–91 as part of a comprehensive study of weasels and their prey in primeval forest (Jedrzejewski et al., 2000). Forest rodent numbers irrupted following the synchronous seeding of oaks, hornbeams, and maples, peaking in 1990 at more than 300 rodents per ha and then crashing. The weasels were active predominantly during daylight, with a peak around noon. When rodents were abundant on the forest floor, the ranges of five radio tracked male weasels ranged from 11 to 37 ha, but after the rodents disappeared, the two different males radio tracked after the crash in rodent densities covered 117 and 216 ha (Jedrzejewska & Jedrzejewski, 1998; Jedrzejewski et al., 1995). So weasels respond to resource pulsing as do stoats, but faster. Trapping regimes should not assume home range sizes are constant regardless of prey resources.

In Wytham Wood, Oxford University's research estate (Sect. 2.1), Tom Tew was the first to try radio-collaring English weasels, three females and seven males. He followed them for a few days each in 1991–92, on the farmland the other side of the estate from my old study area in Marley Wood (MacDonald et al., 2004). As in the Wytham weasels I observed decades earlier, and like those observed in Poland, Tom's radio-collared weasels were active only in daytime. Understandably, radio-telemetry suggested that Wytham weasels had much larger home ranges than could be found by live trapping over a shorter period. The four adult males (mean home range 113 ha) and two nonbreeding females (28 ha), lived in the network of linear habitats criss-crossing the farmland adjacent to the woods. Successively tracked males occupied overlapping ranges calculated as both minimum convex polygons and restricted polygons confirming that every vacated home range was quickly taken over by others. Linear strip home ranges were defined as 10 m wide strips following landscape features used by weasels, showing that none moved far from the cover provided by hedges and ditches.

Another study led by Xaviar Lambin in Kielder Forest (Northumberland, UK) used radio-tracking to follow the movements of eight male weasels hunting field voles in thick grassland (Brandt & Lambin, 2007). They agreed that weasels were more active in daylight, when voles were more active and easy to catch.

9.3 Radio Telemetry 2: Stoats and Ferrets in New Zealand

Cost-effective pest management work depends on having reliable prior information on the home ranges and normal activity routines and travel capabilities of target species. The arrival of radio-telemetry technology in New Zealand in the 1990s provided the first opportunities to work out these critical details for stoats and ferrets.

Home Range of Stoats In the Eglinton Valley, Elaine Murphy and John Dowding ran the first study to equip stoats with radio-collars. They showed clearly how stoats adjusted their home range sizes to the abundance of prey (Murphy & Dowding, 1994, 1995). During two consecutive summers, wild house mouse populations boomed and then busted (Sect. 5.3). When mice were everywhere in 1990–1991, three male stoats had mean home ranges of 93 ha, and four females, 69 ha, and they were often active during the day. In the same area in the next summer, when mice were scarce, the equivalent numbers were 204 ha for four males and 124 ha for five females. The same technique was used later in Pureora Forest to monitor the daily survival of stoats after a 1080 drop (Fig. 8.1).

Incidentally, this study supplied an intriguing comment on the use by stoats of road-killed carrion. The Eglinton Valley carries heavy traffic along the main tourist road to Milford Sound. It was one of our original National Parks survey areas (Sect. 4.3), where we had found possum remains in up to 14% of 212 stoat guts. Several radio-collared male stoats actively preferred to hunt along the road, and were often found denning near road-killed possums. One managed to drag a possum (a deadweight of 5–10 times its own body mass) to a nearby den. Some of them left their escape from oncoming traffic a fraction too late (Fig. 4.3).

Similar results were recorded by three more radio-tracking studies that followed in quick succession, which illustrated how the results could be affected by habitat and intra-guild competition. One team completed two studies in southern beech forest. First, when mice were scarce, they found that the home ranges of stoats (four males averaging 223 ha, and of seven females, 94 ha) were evenly spaced and overlapped slightly at a few extreme data points. Then they observed the home ranges of six male stoats in coastal scrub/grassland averaged 133 ha, and of two females 83 ha, which hunted both by day and by night (Alterio, 1998; Alterio & Moller, 1997; Moller & Alterio, 1999).

The third study was run in two podocarp forests in south Westland, a habitat completely different from either beech forest or coastal scrub. There, 27 stoats were fitted with radio transmitters and tracked from July 1997 to May 1998 (Miller et al., 2001). This large project collected extensive data on 19 stoats, and analysed them by three different methods. The mean size of the home ranges across all seasons in males was 210 ha, and for females, 89 ha when calculated by the minimum convex polygon method, most useful for comparing with other studies. Males extended their ranges during the breeding season (256 ha) compared with the nonbreeding seasons (149 ha), but females did not. Home ranges apparently overlapped within and between sexes in all seasons, most extensively between male and female ranges.

Supplementing Trapping Data Trap avoidance or over-wide trap spacing can protect the stoats surviving in managed mainland populations, and underestimates their distribution and activity. So all these studies are useful to managers deciding how to space their traps. Radio-tracking can also be key to confirming the effects of secondary poisoning after a 1080 operation (Figs. 8.1 and 8.2) and for finding stoat dens. A comprehensive study of the ecology of stoats in the Tasman Valley in 2000–2002 (Dowding & Elliott, 2003), and confirmation of secondary poisoning on stoats after a 1080 drop (Chap. 8) could not have been done without prior radio-collaring of the resident stoats.

Like stoats, ferrets are also very cautious about approaching human devices. Three successive studies of radio-collared ferrets on North Island pastoral land using an experimental smart bait dispenser (Sect. 9.5), recorded data proving that wild animals often take longer to find or enter a trap than it remains available, or actively avoid the traps that they encounter, or refuse to take bait. The alarming implication is that trap avoidance in wide-ranging mustelids such as ferrets and stoats is probably commonplace. That implies a level of trapping inefficiency that should be of serious concern to conventional pest control programmes, and helps explain why short-term or seasonal control of invasive mustelids is often unable to reduce the densities of target species (King et al., 2009).

9.4 Remote Cameras

In 1993 a new technique of monitoring birds' nests by time-lapse infra-red video cameras began in Pureora Forest (Chap. 6). The tapes confirmed the previously unproven suspicions that possums did not merely compete with kōkako for food, but actively raided kokako nests and destroyed both eggs and chicks. The landmark paper (Brown et al., 1993) published in *Notornis* gave conservationists their biggest and most unpleasant surprise in years. It provided unarguable evidence that possums are the only invasive pests that cause significant direct damage *both* to the forest vegetation *and* to the wildlife of New Zealand.

Carrying delicate video equipment and heavy car batteries around the forest was exhausting, and field staff continued to do it only because the results were so revealing. Lucky for them, hunters wanting to know where to find deer soon developed a deep-pocketed international market driving development of much smaller devices now known as trail cameras or camera traps. These are compact, robust, relatively inexpensive, commercially made and widely available enough to make remote monitoring of bird nests routine.

Disadvantages The technology of detecting and identifying nest predators has revolutionised our understanding of predator-prey relationships in New Zealand. The only issue to cloud this encouraging prospect is the problem of varying detectability. Small and fast-moving animals—like stoats and ferrets –can trigger a camera but be out of range by the time the picture is taken. According to comprehensive comparison trials, some cameras are better at waking up in time than are others (Driessen et al., 2017). Hence standardised reporting of remote camera data (make, model, focal length, reaction time) is needed to enable greater ease of study replication, and thence to improved wildlife research and management outcomes (Meek et al., 2014).

The chances of any kind of camera detecting a small animal are increased if the target can be persuaded to stay close for a few seconds longer. Brandon Breedt had 32 cameras mounted on trees or fence posts, pointing vertically downwards at an affixed rabbit scent lure, which made a foraging mustelid pause for just long enough to let the camera take a picture (Fig. 9.2). This approach worked for all species of small mammals common in each habitat sampled (Breedt & King, 2021).

Fig. 9.2 A trail camera set up on a post, focussed down on the ground, captures an image of a stoat just leaving a camera site after investigating a rabbit scent lure (white package) fixed to the base of the post. The bottom of the image records the details of the visit: temperature 71 °F (22 °C), date 24/03/16, time 16:32:26, and ID:0024. (Photo by Brandon Breedt)

Advantages Trail cameras set to record short videos by day or night can reveal stoat behaviour that might not otherwise be believed. For example, in two high alpine areas of the Southern Alps over three breeding seasons, 2012–2015, Department of Conservation staff monitored 32 rock wren (*Xenicus gilviventris*) nests using motion activated infrared (IR) cameras. Stoats could certainly see the IR lights, but simply ignored them (Sect. 10.2). The behaviour of stoats seen preying upon 13 of these nests was analysed (Sect. 14.2.3). Stoats showed remarkable agility in climbing or jumping across wide gaps to reach nests in apparently inaccessible positions. Trail cameras are also very important in monitoring traps and bait stations, revealing how often animals refuse to enter.

The ultimate challenge in pest spotting is to detect the arrival of a single colonizing individual in a large area devoid of them, such as a pest-free offshore island, or surviving an eradication there. Twenty years ago, footprint tracking was the cheapest and simplest method of remote monitoring. Choquenot (2001) constructed a model predicting how many detection units would be needed to spot the presence of a single stoat on an area of 10,000 ha (100 km^2) with 75% confidence.

Nowadays, trail cameras would do the job much more efficiently (Roilo et al., 2024), but the parameters of Choquenot's model roughly fit either device. He estimated that some 350 devices would be needed to detect a single invading stoat on 10,000 ha (assuming each tunnel sampled 1.5 ha, a probability of 0.7 that a stoat encountering a device will be recorded by it, and that home ranges average 50 ha with 20% overlap). Operating that many devices over such a large area would be prohibitively expensive in equipment and staff costs. If five stoats were present, only about 50 devices would be needed to detect at least one of them with the same confidence. That would be more achievable, but five stoats reaching the island would amount to a serious incursion that would become obvious from sightings, scats and prey remains. Furthermore, a 0.7 probability of detection is far too optimistic (Chap. 11).

Monitoring Pest-Exclusion Fences Remote cameras have found another important role in 24/7 monitoring the security of fenced sanctuaries. The technically advanced Xcluder TM pest-proof exclusion fence developed in New Zealand (Day & MacGibbon, 2007) has a tall framework covered in fine mesh (7.1 mm × 40 mm) surmounted by a rolled hood enclosing a gutter accessible only from the outside. Animals climbing the fence could reach the inside of the rolled hood and run along inside the gutter, but could not climb onto the top of the fence.

In one experiment, a series of holes in the fence simulating breaches due to a falling tree or a flood scour were each backed by a cage and monitored by a video camera (Connolly et al., 2009). Other cameras watched the inside of the gutter atop the fence. All the small mammals present in the surrounding farmland routinely patrolled the fence searching for a way in (Fig. 9.3). Sightings were analysed by standard GLM and ANOVA statistics, from which a model aimed to describe the chances of a given species of pest gaining entry.

Among 788 sightings recorded in summer, and 428 in winter, only two stoats were seen by the outside cameras. Ship rats were more often detected on the ground than were stoats (83 in summer, 42 in winter) and they also travelled and foraged on insects within the gutter inside the rolled hood on top of the fence (Fig. 13.1). The most common sightings were of possums, cats or mice. Taking all sightings together, the conclusion was that any breach in the fence will be found and exploited within 24 h, which means that managers have to respond very rapidly when the remote alarm system installed along the top of the fence goes off.

9.5 A Smart Bait Dispenser

Whole eggs have been for years the most highly favoured bait in traps because they don't go bad as fast as raw meat does, but there was still the problem that, once the egg was taken, someone had to go out and replace it. In 2000–2006 a bioengineering research group at Ruakura, led by engineer Rod McDonald, aimed to test his idea of putting a couple of hundred doses of liquid egg-mayonnaise bait in an aerosol can set in the tunnel of a new gadget called the Scentinel. A weight-sensitive trigger could go on offering a few drops of bait to successive visitors without needing attention for months at a time. It was important to know how animals would respond to such a novel bait, so those entering the tunnels were monitored by an internal camera. Over the next 5 years, the team conducted a long series of prototypes and field

Fig. 9.3 (**a**) An experimental hole in the Xcluder TM predator-proof fence surrounding Maungatautari Sanctuary Mountain was monitored by remote cameras and backed by cage inside. (Photo by Author). (**b**) A stoat found the hole, but could not get past the cage. Climbing up the outside of the fence led only to an overhanging slippery metal hood offering no further access. (Photo by Trevor Connolly)

trials advancing the design, all but the last using non-toxic bait. The Scentinel provided many revealing insights into how animals react to human devices.

Field Trials The first prototype was put through its first field trial in Fiordland beech forest in December–February of 2000/01 (Purdey et al., 2004). It was the summer following a huge post-seedfall mouse irruption in 1999–2000 and the subsequent population crash, so the forest was full of hungry young stoats competing for the last few mice.

Over 10 days in November 2000, 25 live stoats were caught, radio-collared and released in the Borland and Grebe Valleys. The Scentinels were set out along a single trap line, and their cameras were there to monitor bait acceptance, not home ranges. But they told us how far stoats ranged along the line of bait stations, and that they loved the egg-mayonnaise mixture.

A new project in 2004 laid out ten Scentinels in pastureland, spaced 2 km apart, and equipped 13 locally resident ferrets with radiocollars . Four of the Sentinels were accompanied by a monitoring station tuned to the frequencies emitted by known radio-collared ferrets. Only six of the 13 ferrets available approached to within 8 m of the four monitoring stations, and the Scentinel's internal cameras detected only three of the six entering its tunnel over 8 days of observation. At the end of the study, we were required to remove all the 15 ferrets we had released on the 6000 ha study area. Five of them eluded recapture , although their radiocollars reported that all remained present (King et al., 2009).

The next trial in 2005 set out 24 Scentinels in a grid at 1 per 100 ha of Waikato pastoral land. From 11 February to 29 April, over 1718 trap nights, 1559 visits by small mammals were recorded, including 198 by ferrets. By the end of the 4th week, ferrets had been detected at 17 of the 24 sites. Removal of ferrets from the study area by contractors began during the 5th week. Site occupancy analyses (Sect. 11.2) confirmed significant declines in the number of ferret visits recorded per day and the number of sites visited (King et al., 2007a), and that ferrets preferred egg-oil mixture to homogenised sheep brain tissue (King et al., 2007b).

Finally, the survival of 23 radio-collared ferrets was monitored before, during and after a 5-week field experiment in 2006, using toxic bait deployed in 20 Scentinels distributed over 2554 ha of dairy pasture. This latest version (Mark 6) used a 'teaser' non-toxic bait dribbled outside the tunnel, automatically renewed daily. Eight ferrets entered a Scentinel and took a lethal bait: four entered but did not take the bait; two did not visit but were killed by secondary poisoning; and nine ignored them. After the experiment, intensive live trapping guided by repeated radio-location surveys retrieved only two of 13 collared ferrets that were definitely still alive on the study area (King et al., 2009). That last and most crucial field trial of our 6-year programme in mid 2006 showed that tiny doses of 1080 could be selectively delivered to radio-collared ferrets over 25 ha, but that not all local ferrets would take it within the time allowed.

Downstream Benefits Much of the Scentinel programme was funded by the Animal Health Board (now called Ospri), keen to know how far ferrets move spreading bovine tuberculosis to cattle. But just as the Scentinel proved how much useful field data it could collect from small mammals generally, the funding ran out. The programme was cancelled, and the Scentinel, too far ahead of its time, was never commercialised. Nonetheless, our published reports of the field experiments remain available to inform others using different technology to address the same questions.

Many years later, Rod McDonald's IP for the Scentinel mechanism was acquired by other players, to support the development of some new, unrelated downstream projects capable of attracting new funding. One, called the Spitfire, is a tunnel containing a self-resetting delivery system that sprays toxin, such as PAPP (Chap. 8), onto the belly fur passing of stoats or rats, relying on their natural instinct to groom themselves clean (Blackie et al., 2014; Murphy et al., 2023). Another one has taken the concept of an automatic self-resetting liquid egg/mayonnaise bait dispenser and developed it into a stand-alone lure dispenser that can be placed beside any trap or tunnel, refreshed at the rate of 0.15 ml every night. These gadgets are now commercialised under the name MotoLures (https://zip.org.nz/products-list/motolure), and they increase detectability by rewarding repeat visits (Cook & Mulgan, 2022).

Any review of current technology in a fast-moving field will soon be outdated, and speculations about future devices are pointless. Rapid advances in image processing, assisted by AI and other third-party tools, will be able to classify images from trail and thermal cameras at great speed—as required for managing pest incursion response operations on offshore islands and fenced sanctuaries (Gillies, 2025). The prospects are fascinating. But this chapter can at least encourage further progress by showing how every previous laborious advance in how data are collected and handled can open up exiting new fields of information.

References

Alterio, N. (1998). Spring home range, spatial organisation and activity of stoats *Mustela erminea* in a South Island *Nothofagus* forest, New Zealand. *Ecography, 21*, 18–24.

Alterio, N., & Moller, H. (1997). Daily activity of stoats (*Mustela erminea*), feral ferrets (*Mustela furo*) and feral house cats (*Felis catus*) in coastal grassland, Otago Peninsula, New Zealand. *New Zealand Journal of Ecology, 21*, 89–95.

Blackie, H. M., MacKay, J. W. B., Allen, W. J., Smith, D. H. V., Barrett, B., Whyte, B. I., Murphy, E. C., Ross, J., Shapiro, L., Ogilvie, S., Sam, S., MacMorran, D., Inder, S., & Eason, C. T. (2014). Innovative developments for long-term mammalian pest control. *Pest Management Science, 70*(3), 345–351. https://doi.org/10.1002/ps.3627

Brandt, M. J., & Lambin, X. (2007). Movement patterns of a specialist predator, the weasel exploiting asynchronous cyclic field vole Microtus agrestis populations. *Acta Theriologica, 52*(1), 13–25. https://doi.org/10.1007/bf03194195

Breedt, B., & King, C. M. (2021). Distribution and detectability of mammalian pests in the Waikato Region. *New Zealand Journal of Zoology, 49*, 37–52.

Bridgman, L. J., Innes, J., Gillies, C., Fitzgerald, N. B., Miller, S., & King, C. M. (2013). Do ship rats display predatory behaviour towards house mice? *Animal Behaviour, 86*, 257–268.

Bridgman, L., Innes, J., Gillies, C., Fitzgerald, N., Rohan, M., & King, C. M. (2018). Interactions between ship rats and house mice at Pureora Forest Park. *New Zealand Journal of Zoology, 45*, 238–256. https://doi.org/10.1080/03014223.2018.1464477

Brown, K., Innes, J., & Shorten, R. (1993). Evidence that possums prey on and scavenge birds' eggs, birds and mammals. *Notornis, 40*, 169–177.

Brown, K. P., Moller, H., Innes, J., & Alterio, N. (1996). Calibration of tunnel tracking rates to estimate relative abundance of ship rats (*Rattus rattus*) and mice (*Mus musculus*) in a New Zealand forest. *New Zealand Journal of Ecology, 20*, 271–275.

Choquenot, D., Ruscoe, W. A., & Murphy, E. (2001). Colonisation of new areas by stoats: Time to establishment and requirements for detection. *New Zealand Journal of Ecology, 25*, 83–88.

Connolly, T. A., Day, T. D., & King, C. M. (2009). Estimating the potential for reinvasion by mammalian pests through pest-exclusion fencing. *Wildlife Research, 36*, 410–421.

Cook, B., & Mulgan, N. (2022). Targeted mop up and robust response tools can achieve and maintain possum freedom on the mainland. *Animals, 12*, 921.

Day, T., & MacGibbon, R. (2007). Multiple–species exclusion fencing and technology for mainland sites. In G. W. Witmer, W. C. Pitt, & K. A. Fagerstone (Eds.), *Managing vertebrate invasive species: Proceedings of an international symposium* (pp. 418–433). USDA/APHIS/WS, National Wildlife Research Center.

Dowding, J. E., & Elliott, M. J. (2003). *Ecology of stoats in a South Island braided river valley* (Contract report on investigation 3405). New Zealand Department of Conservation.

Driessen, M. M., Jarman, P. J., Troy, S., & Callander, S. (2017). Animal detections vary among commonly used camera trap models. *Wildlife Research, 44*, 291–297.

Gillies, C. A. (2025). Personal communication.

Gillies, C., & Williams, D. (2013). *Using tracking tunnels to monitor rodents and other small mammals* (Version 2.5.2). Department of Conservation.

Harris, H. A., Kelly, D., Innes, J., & Allen, R. B. (2022). Invasive species and thermal squeeze: Distribution of two invasive predators and drivers of ship rat (*Rattus rattus*) invasion in mid-elevation Fuscospora forest. *Biological Invasions, 24*, 2547–2559.

Jedrzejewska, B., & Jedrzejewski, W. (1998). *Predation in vertebrate communities. The Bialowieza Primeval Forest as a case study*. Springer.

Jedrzejewski, W., Jedrzejewska, B., & Szymura, L. (1995). Weasel population response, home range, and predation on rodents in a deciduous forest in Poland. *Ecology, 76*(1), 179–195.

Jedrzejewski, W., Jedrzejewska, B., Zub, K., & Nowakowski, W. K. (2000). Activity patterns of radio-tracked weasels *Mustela nivalis* in Bialowieza National Park (E Poland). *Annales Zoologici Fennici, 37*(3), 161–168.

King, C. M., & Edgar, R. L. (1977). Techniques for trapping and tracking stoats (*Mustela erminea*); a review, and a new system. *New Zealand Journal of Zoology, 4*, 193–212.

King, C. M., McDonald, R. M., Martin, R. D., MacKenzie, D. I., Tempero, G. W., & Holmes, S. J. (2007a). Continuous monitoring of predator control operations at landscape scale. *Ecological Management and Restoration, 8*, 133–139.

King, C. M., McDonald, R. M., Martin, R. D., Tempero, G. W., & Holmes, S. J. (2007b). A field experiment on selective baiting and bait preferences of pest mustelids (Mustela spp.). *International Journal of Pest Management, 53*, 227–235. https://doi.org/10.1080/09670870701376382

King, C. M., McDonald, R. M., Martin, R. D., & Dennis, T. (2009). Why is eradication of invasive mustelids so difficult? *Biological Conservation, 142*(4), 806–816. https://doi.org/10.1016/j.biocon.2008.12.010

King, C. M., Gaukrodger, J. D., & Ritchie, N. A. (2015). *The Drama of conservation: A history of Pureora Forest, New Zealand*. Springer/New Zealand Department of Conservation.

MacDonald, D. W., Tew, T. E., & Todd, I. A. (2004). The ecology of weasels (*Mustela nivalis*) on mixed farmland in southern England. *Biologia, 59*(2), 235–241.

Meek, P. D., Ballard, G., Claridge, A., Kays, R., Moseby, K., O'Brien, T., O'Connell, A., Sanderson, J., Swann, D. E., Tobler, M., & Townsend, S. (2014). Recommended guiding principles for reporting on camera trapping research. *Biodiversity and Conservation, 23*, 2321–2343. https://doi.org/10.1007/s10531-014-0712-8

Miller, C., Elliot, M., & Alterio, N. (2001). Home range of stoats (*Mustela erminea*) in podocarp forest, South Westland, New Zealand: Implications for a control strategy. *Wildlife Research, 28*, 165–172.

Moller, H., & Alterio, N. (1999). Home range and spatial organisation of stoats (*Mustela erminea*), ferrets (*Mustela furo*) and feral house cats (*Felis catus*) on coastal grasslands, Otago Peninsula, New Zealand: Implications for yellow-eyed penguin (*Megadyptes antipodes*) conservation. *New Zealand Journal of Zoology, 26*, 165–174.

Murphy, E. C., & Dowding, J. E. (1994). Range and diet of stoats (*Mustela erminea*) in a New Zealand beech forest. *New Zealand Journal of Ecology, 18*, 11–18.

Murphy, E. C., & Dowding, J. E. (1995). Ecology of the stoat in *Nothofagus* forest: Home range, habitat use and diet at different stages of the beech mast cycle. *New Zealand Journal of Ecology, 19*(2), 97–109.

Murphy, E. C., Clapperton, B. K., Bradfield, P. M. F., & Speed, H. J. (1998). Effects of rat-poisoning operations on abundance and diet of mustelids in New Zealand podocarp forests. *New Zealand Journal of Zoology, 25*, 315–328.

Murphy, E., Agnew, T., Sjoberg, T., Eason, C. T., McMorrran, D., & Ross, J. D. (2023). Developing a new resetting tool for controlling rats. *New Zealand Journal of Ecology, 47*, 3517.

Purdey, D. C., King, C. M., & Lawrence, B. (2004). Age structure, dispersion and diet of a population of stoats (*Mustela erminea*) in southern Fiordland during the decline phase of the beechmast cycle. *New Zealand Journal of Zoology, 31*, 205–225.

Roilo, S., Hofmeester, T. R., Frauendorf, M., Widen, A., & Cord, A. F. (2024). The untapped potential of camera traps for farmland biodiversity monitoring: Current practice and outstanding agroecological questions. *Remote Sensing in Ecology and Conservation*. https://doi.org/10.1002/rse2.426

Sweetapple, P., & Nugent, G. (2011). Chew-track-cards: A multiple-species small mammal detection device. *New Zealand Journal of Ecology, 35*, 153–162.

Walker, S. (2025). Personal communication.

Walker, S., Kemp, J. R., Elliott, G. P., Mosen, C. C., & Innes, J. G. (2019). Spatial patterns and drivers of invasive rodent dynamics in New Zealand forests. *Biological Invasions, 21*, 1627–1642.

Captive Studies 10

Abstract

Close observation of the behaviour of fast-moving mustelids in the field is difficult over any useful length of time. The behaviour of stoats and ferrets adapted to captivity may not be entirely normal, but experimental testing of hypotheses can be done there provided the conditions are sympathetic to their nature. Neophobia (refusal to engage with strange devices) can cause learned aversion to field equipment. It also varies between individuals, so distorts the assumption of equal catchability. Captive stoats, used to a female carer, reacted with especially pronounced body rubbing and anal gland scent marking to the scent of human males introduced into their cage. Remote cameras normally use infra-red (IR) lighting, invisible to humans, for video surveillance. Ferrets trained to respond to a visual stimulus responded with 89% accuracy to a white light (LED); two of five continued to respond at 84% and 78% to a standard IR (870 nm) LED. One could not see a 929 nm LED. Competition within New Zealand's predator guild is mitigated by the environment. Stoats confined to a pen are scared of a direct confrontation with caged cat, but in nature they can avoid each other via different activity schedules.

Keywords

Nephobia · Scentmarking · Infrared (IR) vision · IR nest monitoring · Interference competition · Activity schedules

The Problem Weasels and stoats have rapid metabolic rates and high-energy hunting instincts, so they live their lives at breakneck speed and do not cope well with confinement. But, if provided with five-star accommodation, fed freshly dead natu-

C. M. King, *Stoat in the Dock*, Fascinating Life Sciences,
https://doi.org/10.1007/978-981-96-4562-6_10

ral prey, and kept quiet and very clean, they can adapt for long enough to answer some particular questions.

In the 1880s, Henry Allbones of Brigg, UK and his son Walter, operated a system for transporting live stoats and weasels from Britain to New Zealand, on steam ships that took up to 6 weeks at sea to sail halfway round the world, defying predictions that they would dash themselves to pieces in the cages. They delivered cargoes of hundreds at a time, with in-transit losses seldom exceeding 10% (Sect. 1.2), although other less careful operators lost many more.

Captive colonies of stoats have provided vital documentation of the reproductive cycles that could not be obtained any other way. Philip Wright (1942) confirmed the compulsory delay in implantation that interrupts the reproductive cycle of stoats and longtails (Sect. 4.5), and conducted experimental surgeries confirming the hormonal control of it. A long-running captive colony in Germany added details on mating season, mating behaviour, and litter size (Müller, 1970). In Russia, Ternovsky (1983) maintained a captive-bred population of stoats in Novosibirsk for decades, from which he provided proof of the astonishing precocity of juvenile females (Sect. 4.5).

These papers underline the importance of keeping track of the older literature, because they provided essential background for our understanding of stoat reproductive biology in New Zealand. Known-age skulls from Fritz Frank's long-running colony of weasels proved fundamental to the development of age-determination techniques for both weasels and stoats (Sect. 2.4). More recently, investigations into the prospects of fertility control for New Zealand stoats (Chap. 13) established a captive colony which has also turned out to be useful for other reasons (O'Connor et al., 2004) .

Pet weasels tamed as kits and given the run of a house or very large enclosure can teach a trusted human companion much important information about uninhibited natural behaviour (Drabble, 1973, 1977). Robert Fuller (robertfuller.com) and Andrew Veale (2024) have closely observed and photographed the behaviour of semi-tame stoats over long periods.

But for their wild-caught cousins, close confinement of such high-powered animals is stressful enough to change the natural behaviours of interest. Captive studies have other problems too, such as small sample sizes, but they also allow a specific variable to be controlled whilst keeping others constant. On balance, captive stoats and ferrets can offer some useful insights impossible to document any other way.

10.1 Neophobia

Wild stoats are small, fast-moving animals that seldom stand still for long. A brief glimpse of a brown, tubular form diving for cover can provide only an unpredictable confirmation of their presence. Radio-telemetry has revealed detailed new information about their denning, territorial and ranging behaviour relevant to control operations (Sects. 9.2 and 9.3) but that first requires a capture, then observations mostly from a distance. Any hope of making systematic, controlled direct observations of stoat behaviour at close range is possible only by watching them in captivity.

Lloyd Robbins (2007) designed a set of comfortable cages where stoats could feel safe and at home. The cages were long, wide and low, filled with hay imitating the sort of overgrown ungrazed grassland where wild stoats like to hunt. They made their own tunnels through it, and also readily ran through and scent-marked the plastic pipes supplied as refuges. A detachable external nest box accessible through a sliding door gave them privacy, and when closed, allowed the depositing of bait or cleaning of the runways without disturbing them.

Neophobia is the spontaneous refusal of new food, or fear of approaching novel or apparently threatening objects, situations or spaces. Field results confirm, all too often, that the problem of aversive responses by stoats and ferrets to unfamiliar baits or control devices is very real. For example, time lapse video recorded that 20% of wild stoats did not enter tunnels baited with hens' eggs (Dilks & Lawrence, 2000). Only a single negative experience can establish a learned aversion that will persist for life. Research into the causes of and solutions to bait/device shyness is obviously important to improve control operations in New Zealand, but had never previously been done.

Trap avoidance is an important issue for stoat control operators. For example, one observer describes a battle of wits he had with a stoat that learned, from a single experience, to avoid a further 32 carefully sited traps (Cahn, 1936). Elaine Murphy and John Dowding set a ring of live traps around a den known to contain a litter of young stoats, of which up to four at a time were seen playing around the den entrance, but none was ever caught (Sect. 9.3).

The same applies to ferrets. A Department of Conservation (DOC) licence permitting us to release 30 radio-collared ferrets for a field trial required any survivors to be retrieved and destroyed at the end of the study. In the event, 13 ferrets were still alive on the study area, and radio signals showed where they were, but intensive live trapping guided by repeated radio-location surveys could not catch eleven of the 13 (Sect. 9.5).

Does the presence of human odour have any effect on the behaviour of stoats towards control devices? Time-lapse video cameras set up to observe the reactions of eight stoats to novel baits or objects and to human scents strongly suggested that stoats are not so much neophobic as cautiously curious about novel objects, repeatedly approaching and retreating (Watson & King, 2020). Once the stoats were habituated to using plastic tubes as tunnels, the tubes were used to test their reactions to novel odours, both neutral (goat) and potentially threatening (human). The frequencies of scent-marking behaviours on treated tubes were compared with those from a baseline day that demonstrated the animals' normal behavioural repertoire.

Six of the eight stoats performed significantly more rubbing and anal gland scent marking in response to human odour than to any other treatment. (NB the human odour was collected from used t-shirts, but only from males, because the stoats were used to daily visits from a female carer). It seemed that the stoats were using scent marking to convey agonistic and territorial reactions, perhaps as a substitute for their likely normal reaction in the wild, to flee. The implication, that human scent on traps or tunnels could contribute to neophobic avoidance of control devices, has not yet been formally tested.

10.2 Can Ferrets See Infra-red Camera Lights?

Since the 1990s, remote cameras have been widely used to watch bird nests at night under intra-red lights (Chap. 9). The standard equipment emits infra-red (IR light) at wavelengths between 830 and 880 nm, which are generally invisible to humans. But field staff have always wondered whether animals can see further into the IR spectrum than we can, and if so, does IR lighting actually guide predators toward a nest? Video surveillance is important because it can document behavioural interactions as well as species identification, but the potential risk of IR lights providing beacons for nest predators is alarming. Obviously, it is essential that the IR light used for nocturnal surveillance is undetectable so that it does not influence the behaviour of hunting predators. How far is that is true for ferrets?

Haylie Newbold took five farm-bred male ferrets, accustomed to living in cages, housed them singly and taught them to press a metal bar to receive food reward if they could see a test stimulus (a light emitting diode, LED). They all learned to respond with an average of 89% accuracy to a white LED. When the white light was changed (via a visible red light) to IR (870 nm), two of five ferrets responded strongly at 84% and 78%. One of these could definitely not see further into IR at 920 nm (47% correct response) (Newbold & King, 2009). These results suggested the range of ferret's visual capability but not proof of it, since we could not tell the difference between a ferret failing to report what it could see versus becoming bored with whole business. Nevertheless, video equipment used in the field now mostly uses using 'No-Glow' cameras in the 940 nm wavelength (Gillies, 2025).

Video records from older studies do sometimes show stoats and ferrets looking directly at the IR light, supporting the implications of these results that at least some of them could detect the standard equipment. Further tests would be useful, but unlikely. The ferret training programme was extremely laborious, and time did not permit further exploration of the differences in behaviour of these individuals. The ferret is a declared noxious species in New Zealand, so the 60-page Department of Conservation research permit took 6 months to negotiate, and required that all the ferrets be put down at the end of the study.

10.3 Competition Among Mustelids

In New Zealand, stoats, weasels and ferrets belong to a guild of introduced predators, of which the largest and dominant members are feral cats. Stoats are smaller than both ferrets and cats, so are vulnerable to interference from both, and have to balance the need to hunt against the risk of a hostile encounter while hunting. Weasels are vulnerable to all larger predators, and their compensating foraging advantage in access to small rodent burrows, which protects them in arctic and boreal predator communities, does not apply in New Zealand (King et al., 2017). So weasels are scarce here except where their larger competitors have been effectively suppressed.

Where stoats can be removed more effectively than can weasels, the release of competition helps to swing the balance in favour of the smaller of the two. Weasels have an advantage in slipping through tiny holes under fences. The Brook Waimarama Sanctuary, near Nelson, was invaded by a few weasels some time in 2019. Inside, in the absence of stoats and rats but surrounded by mice, their numbers shot up to 73 weasels trapped inside the fence by August 2020 (Sect. 13.2). Outside fences, traps set too far apart for stoats often miss weasels, which begin to turn up in greater numbers. The informal literature is beginning to report many such cases in New Zealand, but few are yet fully analysed.

Overseas, the evidence is more conclusive. Effective removals of stoats (that happened for different reasons) in England (Sumption & Flowerdew, 1985) and Belarus (Sidorovich & Solovej, 2007) were followed by significantly increased numbers of weasels.

For hunters that can climb well, New Zealand forests hold more food than do the open, hedgeless paddocks that have replaced them. Moreover, cats and/or ferrets are more common in open country than in forest (Keedwell & Brown, 2001), and both can easily beat up a stoat. So, as documented in the Tasman Valley in 1997/98, there is usually a negative correlation between the capture rates of stoats and ferrets (Dowding & Elliott, 2003). So habitat preferences are the important first step in avoiding competition.

To observe the interactions between predatory species more closely, several large outdoor pens were designed to watch the reactions of 16 stoats allowed (separately) free to roam under the eye of a video camera (Garvey et al., 2015). A cat or ferret could be held in a small separate cage inside the enclosure, from which it could glare at the stoat but not touch it. When half of a stoat's daily meal was placed too close to a cat or ferret, it would assess that area as risky (even though it was not), and take less food from there than from the other half of the meal placed in front of an empty cage (Fig. 10.1).

In nature, stoats seldom confront ferrets or cats, preferring to adjust their activity schedules to prevent unexpected encounters. In pasture land and other areas where ferrets and stoats might meet, ferrets are dominant over stoats in competition for prey, so stoats avoid ferrets by coming out during the day, while ferrets are most active at night (Garvey et al., 2022). But stoats still actively monitor ferret odour as a way of keeping track of where danger may lie.

Because they communicate with each other by scent marking, pest managers who can understand these chemical messages can create opportunities to listen in to the conversation. In a follow-up field study, the number of stoats seen by cameras approaching a ferret-scent lure increased fourfold compared with a simple food-based lure. (Perhaps this is a similar response to that observed in captive stoats reacting to human scent in their cage). Scents that carry significant messages may be much more effective as lures than standard food baits, especially when natural food is abundant. Chemical cues make an uncertain world more predictable for

Fig. 10.1 A captive stoat on high alert, in a pen with a caged cat. (Photo by Patrick Garvey)

species that are solitary, inhabit complex habitats, or are nocturnal (Garvey et al., 2017).

References

Cahn, A. R. (1936). A weasel learns by experience. *Journal of Mammalogy, 17*, 386.

Dilks, P. J., & Lawrence, B. (2000). The use of poison eggs for the control of stoats. *New Zealand Journal of Zoology, 27*, 173–182.

Dowding, J. E., & Elliott, M. J. (2003). *Ecology of stoats in a South Island braided river valley* (Contract report on investigation 3405). New Zealand Department of Conservation.

Drabble, P. (1973). Life with a stoat, and a weasel. *The Field* (6 December 1973), pp. 1508–1509.

Drabble, P. (1977). *A weasel in my meatsafe*. Michael Joseph.

Garvey, P. M., Glen, A. S., & Pech, R. P. (2015). Foraging ermine avoid risk: Behavioural responses of a mesopredator to its interspecific competitors in a mammalian guild. *Biological Invasions, 17*, 1771–1783.

Garvey, P. M., Glen, A. S., Clout, M. N., Wyse, S. V., Nichols, M., & Pech, R. P. (2017). Exploiting interspecific olfactory communication to monitor predators. *Ecological Applications, 27*, 389–402.

Garvey, P. M., Glen, A. S., Clout, M. N., Nichols, M., & Pech, R. (2022). Niche partitioning in a guild of invasive mammalian predators. *Ecological Applications, 32*, e2566.

Gillies, C. A. (2025). Personal communication.

Keedwell, R. J., & Brown, K. P. (2001). Relative abundance of mammalian predators in the upper Waitaki Basin, South Island, New Zealand. *New Zealand Journal of Zoology, 28*(1), 31–38.

King, C. M., Norbury, G., & Veale, A. J. (2017). Small mustelids in New Zealand: Invasion ecology in a different world. In D. W. Macdonald, C. Newman, & L. A. Harrington (Eds.), *The biology and conservation of musteloids* (pp. 257–277). Oxford University Press.

Müller, H. (1970). Beiträge zur Biologie des Hermelins, *Mustela erminea* Linné, 1758. *Saugetierkundliche Mitteilungen, 18*(4), 293–380.

Newbold, H. G., & King, C. M. (2009). Can a predator see 'invisible' light? Infrared vision in ferrets (Mustela furo). *Wildlife Research, 36*(4), 309–318. https://doi.org/10.1071/wr08083

O'Connor, C. E., Turner, J., Scobie, S., & Duckworth, J. (2004). Stoat reproductive biology. *Science for Conservation, 268*, 24.

Robbins, L., Clapperton, B. K., Tallentire, K., & Porter, R. E. R. (2007). Testing the attractiveness and palatability of stoat lures and baits with various sensory cues. *DOC Research and Development Series, 256*, 1–22.

Sidorovich, V., & Solovej, I. A. (2007). The stoat *Mustela erminea* population decline in northern Belarus and its consequences for weasels *Mustela nivalis*. *New Zealand Journal of Zoology, 34*(1), 9–23.

Sumption, K. J., & Flowerdew, J. R. (1985). The ecological effects of the decline in rabbits (*Oryctolagus cuniculus* L.) due to myxomatosis. *Mammal Review, 15*, 151–186.

Ternovsky, D. V. (1983). The biology of reproduction and development of the stoat *Mustela erminea* (Carnivora, Mustelidae). *Zoologischesky Zhurnal, 62*(7), 1097–1105.

Veale, A. J. (2024). Personal communication.

Watson, S. W., & King, C. M. (2020). Reactions of stoats (*Mustela erminea*) to male human scent. *New Zealand Journal of Zoology, 48*, 159–165.

Wright, P. L. (1942). Delayed implantation in the long-tailed weasel (*Mustela frenata*), the short-tailed weasel (*Mustela cicognani*), and the marten (*Martes americana*). *Anatomical Record, 83*, 341–353.

Advanced Statistics

11

Abstract

Field data from standardised surveys of stoats have been available for decades, but simple descriptive approaches and limited computer capabilities once hampered answering some critical questions. Now, new statistical approaches, run on ever-more powerful computers, can test hypotheses related to those questions. The probability of stoat capture is low, especially on widely spaced trap lines and when mice are not abundant. Stoat populations are patchy, and personality and experience make some individuals especially unlikely to enter a trap. Both of these factors violate essential assumptions of uniform or normal distribution and of equal detectabilities. Standard models of the population dynamics based on capture records can analyse the impacts of trapping on stoat populations. A more sophisticated model for managing unstable populations of stoats in beech forests (Chap. 5) from data with access to variations in age-specific mortality and fertility through the masting cycle can allow for natural mortality. Stoats eat fewer rodents per head than assumed, and cannot prevent post-seedfall rodent irruptions, depress their peak numbers, or delay the next irruption set off by the independent processes controlling mast seeding. Whether removing stoats releases higher numbers of rats or not varies with complex geographic patterns of rodent dynamics.

Keywords

Probability of capture · Occupancy analyses · Trapping impact · Apparent cull · Mesopredator release · Functional response

C. M. King, *Stoat in the Dock*, Fascinating Life Sciences,
https://doi.org/10.1007/978-981-96-4562-6_11

The increasing creativity of hypothesis generation tested with new statistical analyses can lead to exciting new insights into the lives of familiar animals. The main problem is that testing a hypothesis requires large samples of clean, relevant data. Research protocols must include repeated and uniform sampling, free of collection bias and unsupported assumptions. These requirements are especially hard to meet if the subjects of the research are, like stoats and ferrets, small, short-lived, fast moving and live at low density widely scattered across a landscape.

As any military strategist knows, to win a battle one must first understand the enemy. So pest managers need abundant data on the life cycles and reproductive capabilities of their target in a given context before they can develop ways to disrupt its population dynamics more effectively and efficiently. Hence, Parts II and III of this book have described a couple of decades of simple survey work documenting how the basic biology of stoats in New Zealand varies with habitat.

The tools available in the 1970s and 80s to analyse early data collections were the traditional ones of kill traps, carcass dissections, and gut and gonad analyses, histological sectioning and skull measurements, all probed with conventional statistics. Such simple approaches were useful to document the extraordinary lives and resilience of stoats, and their ability to adapt to almost any threat to their survival and reproductive success (Chap. 4).

Those early surveys told us a lot about the characteristics of stoats as targets of conservation management, but did not answer questions about how best to reduce their population densities, because sampling for research did not remove individuals above a level equivalent to sustainable harvesting. But those surveys laid the foundations for modelling techniques that could generate hypotheses not obvious from the data alone. Now, with greater computer power, biologists can use more complex models that combine trap and camera survey data with associated prey density and environmental variables, and to investigate a single variable while keeping all others constant.

11.1 Probability of Capture: Live Traps

Estimating presence or absence, home ranges or population densities of live animals first depended on live capture-mark-recapture (CMR) methods. Unfortunately, live trapping mustelids is often unrewarding, and not representative of the population (Sects. 2.1, 3.1 and 9.3). Where a stoat has been caught once, it must at least have been present there, but empty traps do not necessarily mean that stoats are absent. Worse, the CMR method assumes that marked and unmarked animals are equally liable to be caught, which is often not true. New statistical techniques can now interrogate trapping records to learn whether trapping is representative, and then accommodate unequal catchability and produce estimates not possible before, provided the data sets are large enough.

Gaps in Field Data A simple frequency distribution of live-trapping data (so many individuals caught once, so many different ones caught twice, so many three times, etc.) can be used to infer the presence of individuals in the population that are never caught, which are often a substantial proportion of the local population. Worse, females have small home ranges that can easily fit between widely spaced traps.

Stoats that have been captured once are less likely to be caught again. A spin-off study by Italian mathematician Francesco Bartolucci re-examined my live-trapping records from Fiordland (King & McMillan, 1982), and concluded that a capture is a bad experience for stoats, some of which quickly learn how to avoid future captures (Bartolucci & Pennoni, 2007). Many marked stoats simply moved out of the trapping area, or were not even present during the first time the traps were set.

What proportion of a local population of stoats is caught and marked in a live-trapping survey? Attempts to answer this question are usually frustrated, because of the huge labour of operating enough live traps to get a respectable sample. In January 1980, following a major beech seedfall in 1979, a mass influx of young stoats in Fiordland offered me a chance to organise a CMR study (Sect. 5.1). In early summer, 134 stoats were caught, tagged and released (113 young of the year, many still distinctly smaller than adults, and with no signs of breeding) plus 19 adults and two of unknown maturity. Of one group of 21 stoats, tagged on or before January 15 and known to be present in the same area on or after January 25 1980, nine were not caught on any of the 7 days on which the traps were set between January 15 and 25. In other words, about 45% of the 21 stoats known to be alive and present nearby were not recaptured in a whole week's trapping (King & McMillan, 1982).

A lucky meeting at a conference with CMR expert Stephen Davis led to a new analysis looking more closely at why that recapture rate was so low. The field records were put through his model, showing that, for a short period of very high numbers of mice in January, the estimated probability of a stoat being captured for the first time was only about 17% per day (King et al., 2003). The probability of recapture was higher for adult males than for adult females, and Bartolucci's independent analysis of the same data set confirmed individual variation in trap response, and consistent trap avoidance. The probability of recapture was even lower for the young of either sex, which were more likely to have dispersed away.

Does it matter that live-trapping does not sample a local population very accurately, when conservation managers don't use live traps? Yes, because characteristics of populations that cause stoats to avoid traps are largely universal for both live and kill trapping. In addition, biased early estimates of home range areas and capture probability still influence control operations. Most concerning is evidence that females are more resistant to trapping, because they are the most important targets for control. All female stoats of all ages are already fertilised (Sect. 4.5), so any one of them can still establish the next generation even if every male in the area is killed. Moreover, many non-residents will pass through a trapping area unseen and uncounted.

This general conclusion on variable trapability is consistent with field observations over a longer period and using completely different technology. One study monitored the responses of stoats to bait stations using miniature video cameras placed inside the tunnels. Over a period of 5 weeks of continuous observation, the

Fig. 11.1 A stoat peers into the entrance of a tunnel before deciding whether to enter. (Photo by Patrick Garvey)

cameras filmed 45 occasions on which a stoat approached the entrance (Fig. 11.1), but on eight of these (18%) it did not enter (Dilks & Lawrence, 2000).

Effects of Prey Density More important, the probability of capture is strongly influenced by the density of preferred prey. In a southern beech forest in Nelson province, after a heavy seedfall in autumn 1995, mice and rats were very abundant at first. As they disappeared, stoat density estimated from arrays of Edgar live traps (Sect. 3.1) dropped (from 4.2/km^2 to 2.5/km^2), but their trappability increased (Alterio et al., 1999).

Traps set in a transect cannot sample all home ranges equally. It may be that some individuals were simply not close to a trap during the days those traps were available. Perhaps more important, different categories of both sexes vary a great deal in the way they react to traps. Some, often the old and more experienced individuals, tend to be cautious and more shy of traps than young stoats, especially the newly independent and bolder ones. As the wartime RAF instructors used to say, there are bold pilots, and old pilots, but no old, bold pilots. Trapping selects those individuals that are most likely to engage with human devices, missing those less likely to engage. Trap evasion may be more about personality than about learning (Johnstone & Garvey, 2023; Johnstone et al., 2024).

On top of that, field methods that place traps evenly across a landscape inadvertently choose trap sites in places that stoats never visit, so provide fewer opportunities for some stoats to be captured than others. On conventional trap lines, individuals differ in their exposure to and engagement with traps. One solution to this problem is offered by Spatially Explicit Capture–Recapture (SECR), which models the spatial relationships between animals and detectors (Efford & Fewster, 2012). Vattiato et al. (2023) offer a comprehensive review of both spatially explicit capture-recapture and home-range studies reporting estimates for stoats, and for nine other invasive mammal pests. They show how two commonly applied spatial detectability parameters vary under different biological, environmental, and surveillance conditions, with implications for understanding pest behaviour and efficacy of different surveillance methods. They discuss the correlation, if any, between detectability and habitat, season, surveillance device, and sex, as well as collating information on lure, trap grid spacing, masting, and population status.

Site Occupancy Analyses The logic of occupancy analyses is similar to that of capture-mark-recapture, but is applied to sites rather than to individuals—provided that the cameras are so far apart that the same individual cannot appear at more than one site. It estimates (1) the probability that at least some members of a species are present (occupying) a monitoring site, and (2) estimates the probability that a member of that species will be detected, given that at least one individual is there. From these 'capture' data (trapping, remote camera, direct observation and other data), occupancy analysis can detect uneven distributions of local populations and trap evasion.

Systematic arrays of cameras producing data amenable to occupancy analyses can document the distribution and detectability of many pest species at once (Chap. 9). The method requires time-stamped 24/7 data from multiple site visits recorded over multiple days over at least 2 weeks at a time—a short enough period to limit local population change, but long enough to offer many capture possibilities.

Occupancy analysis is a suitable tool to survey the distribution of small mammals over areas much too large for CMR. Over 12 months starting from March 2016, cameras set across three representative habitats in the Waikato Region provided data to make occupancy estimates for five target species over each of four seasons (Breedt & King, 2021). Only 10 of the 60 estimates (17%) returned probabilities of detection greater than 50% for that species in that habitat at that season, given that at least one individual was present.

Individuals of all the species monitored (possum hedgehog, ship rat, and house mouse, plus the three mustelid species treated as a single group) were present on pastoral land, even those that prefer the remaining patches of natural cover but need to travel between them. In New Zealand, all these species count as invasive predators responsible for various degrees of damage to native fauna (Chap. 1).

The cameras logged the presence of at least one individual of at least one of the three mustelid species in every habitat (Fig. 9.2), and in every season, but at probabilities of detection never exceeding 4% in pasture or indigenous forest, and up to 8% in pine forest. Mustelids occupied a larger estimated proportion of the entire Waikato Region (corrected for habitat coverage) than any of the other pest species listed independently. Such figures are cause for concern among regional pest managers.

At such low detectabilities, occupancy estimates have such huge confidence intervals that they cannot provide precise information on species distribution. However, while rats and stoats are much less common in pasture than in forest, there is much more pasture available (almost 40% of the land area), therefore a larger total number of rats and stoats living in (or at least passing through) pasture than in forest. One study had already confirmed that rats can reinvade forest patches cleared of pests from across surrounding pastures, and do so much faster than expected (King et al., 2011).

11.2 Probability of Capture: Kill Traps

Prey Density and Control Efficiency A perpetual and important question is whether, if stoats are less interested in traps when live mice are abundant, over what range of rodent densities might it operate? And if the effect is real, does that imply less effective stoat population control during the period of critical danger for birds (Sect. 5.4)?

One way to think about this is to construct a descriptive model combining two pre-existing datasets on mouse and stoat abundance derived from routine kill-trapping (King & White, 2004). It could be used to explore the potential range of variation in density of prey that might affect capture probability of stoats through the masting cycle. The older data set spanned two beechmast cycles in northern Fiordland in 1972–80 (Sect. 5.3). The more recent one came from a study of stoats and mice during the aftermath of a huge, unprecedented seedfall in November 1999 in southern Fiordland (Sect. 9.5).

The model results suggested that more stoats were caught when more mice were available, but only up to mouse abundance indices around 25% trap success, typical of an average mast year. Trapability did not decrease significantly until mouse indices exceeded 60%, which happened only once, during a rare superabundant year for mice in 2000/2001.

Traps set on a transect line must not be set farther apart than the width of the average home range. Early field experiments in southern New Zealand showed that close-set traps (100 m between sites) catch a higher proportion of the locally resident stoats, including many females that have smaller home ranges (Sect. 4.7), but closely-spaced traps also cost much more time and effort to operate. Intensive radio-tracking work suggested that the widest trap spacing for stoats, that minimizes effort and still puts at least one trap in every home range, is less than 1 km between sites in grassland and 900 m in podocarp forest (Sect. 9.3).

Closer spacings increase both the work and the number of traps per home range, which in turn increase the probability of putting a trap in the right place. To get five traps in every female's range, spacing must be cut down to about 250 m. The dilemma is that the workload for trap visits by field staff then drastically reduces the trapped area, and hence the number of residents sampled, to fewer than needed for statistical analyses.

Monitoring Control Operations The standard method of monitoring the success of pest control operations is to calculate the Residual Trap Catch Index. RTC indices applied before and after a control operation can estimate the percentage kill rate with associated confidence intervals (Glen, 2014), but cannot detect changes in the detectability of the survivors. By contrast, occupancy analyses are particularly useful to monitor pest control operations, because kill trap data can avoid the problem of ensuring that a single animal cannot be recorded twice.

Occupancy analysis was used to monitor a ferret removal programme on 2400 ha of pastoral farmland over 11 weeks in 2005 (King et al., 2007). For the first 4 weeks, the probability of occupancy and of detection of undisturbed ferrets was recorded at 24 sites, 1 km apart. Over that time the probability of a ferret visiting one of the sites averaged 0.52 (range 0.31–0.73) and the probability of detection was 0.223. Professional contractors starting in the 5th week halved the site occupancy of ferrets to 0.25 (0.07–0.42) and the probability of detecting the survivors averaged 0.15 (Sect. 9.5).

11.3 Revisiting Historic Survey Data

Traditional indices of pest abundance counted only captures per unit effort. For decades, that has been the only practicable means of monitoring populations over the long term. Those old calculations are a still a valuable source of historical data, but the index assumed that the target populations were uniformly distributed, and that all individuals are equally detectable. Neither is always true, if ever. Consequently, if traps failed to catch any of the target species, biologists could not tell whether the animals were present but untrappable, or absent altogether.

New Information from Old Data Statistical number-crunching of long-term trapping records offers one simple way to search for possible associations among identifiable landscape-scale environmental variables and capture success of stoats or ship rats. One such study found such huge variation in results that no firm conclusion was possible beyond what experienced trappers already knew (Christie et al., 2009). For example, stoats were more often caught in a trap that had previously held a rat, or if a stoat had been caught in another trap nearby. Nothing beats local knowledge and field craft—except new statistical techniques.

The New Zealand biological literature includes many decades of extensive capture and tracking records published with only simple analyses. Where these primary datasets are still available, they can be retrieved and re-analysed with new techniques such as occupancy analyses. Sources of bias can be managed, and assumptions identified.

Calculating occupancy requires repeated observations of enough monitoring sites over a sufficient number of sites and days. The large datasets derived from years of standardised rodent and stoat monitoring programmes using quarterly kill trapping, all at fixed sites and inspected daily (Chaps. 5 and 6), met that requirement. A thick pile of folders of old standardised paper records offered student Alison Watkins a treasure trove of material to be digitised and then grilled by occupancy methods.

The data covered four representative temperate forest habitats in New Zealand (three in Pureora Forest Park, Chap. 6) and one in beech forest in northern Fiordland (Chap. 5). The results were uniquely informative. Best-fit models generated by occupancy calculations were consistent with the experience of field staff (Watkins et al., 2010b).

1. There was constant or random probability of occupancy for stoats, which are simply unpredictable. There was dynamic equilibrium probability of occupancy for most populations of mice and rats: the chances of animals arriving or leaving a place were roughly balanced, so the probability of occupancy remains roughly the same, though not static. This result is consistent with the conclusion of Christie et al. (2009), that it is hard to define the exact characteristics of the most successful trap sites.
2. There was widespread site-specific variation in probability of detection, especially substantial in rats and correlated with habitat co-variates. This confirms that it is easier to find rats in some habitats than in others, because habitats are patchy, and therefore so is the distribution of pest populations. A different analysis programme specifically focussed on detecting patchiness in the same survey data came to the same conclusion (Watkins et al., 2010a).
3. Detectability correlated with density indices in mice and rats sampled at 50 m intervals over 3 days. This is a reassuring confirmation that changes in the numbers available to be caught affected the density index much more than variation in individual catchability.
4. The detectability of stoats and rats sampled at 3–400 m intervals over 10 days decreased after 6 days, probably due to a local trap-out effect. Longer-term variations in the density index were consistent with observed changes in reproductive parameters and age structure that are known to follow variations in real numbers.

Many statistical tests have to assume that sampling is representative. The layout of routine trapping operations, though not random, could well be representative if traps are spread evenly. The whole population would then be represented validly by relatively small samples from limited areas. Another assumption is that all individuals are equally detectable and evenly distributed across a study site. If levels of detectability are not clumped, then the distribution of detectability would be evenly sampled across the traplines. Experienced trappers familiar with the real animals know that animals do differ in detectability, as well as in many other characteristics. These analyses, however, concluded that violations of the assumptions of uniform distribution and equal detectability, while real, were not fatal. Standardized surveys run consistently over many years do provide information adequate for short-term population assessments, including important details from carcasses that are not available from tracking tunnels or photographs.

Much larger databases are held by the Department of Conservation, documenting decades of tracking tunnel records across the entire country (Sect. 9.1). They have the advantage of enormous geographic coverage, and have been interrogated by biologists familiar with the more advanced techniques now available. Footprint surveys document presence or absence, which is often the only information required for pest monitoring. But traps can offer carcase analyses, which provide DNA and proteins for many analyses that could not be done before. New hypotheses related to physiology, for example energy consumption and expenditure, body composition, health, microbiota, and so on, need physical specimens. The two sources of material are each valuable on their own terms.

11.4 Population Dynamics

Population models based on large data sets required huge amounts of compilation time in the pre-computer era, leading many biologists to fall back on simple flow charts of inputs and outputs through a year, often adapted from existing models used by fishery monitors and game harvest advisers. The earliest flow charts required only basic guesstimates of production and mortality, and the results were used to set quotas or to check for significant over-harvesting of fish or game.

By the late 1960s, when it became possible to estimate the ages of dead weasels, I applied a slightly more useful analysis to a sample of weasels collected from British gamekeepers (Sect. 2.5). But real progress became possible only when easily accessed computers arrived in the early 1980s.

Models of Trapping Impact Stoats and weasels are valued native carnivores in UK, and seemed to be decreasing nationally. Robbie McDonald was the first to apply this exciting new technique to ask why. He collected a sample of 822 stoats and 458 weasels from 25 widely distributed British game estates, and used calculations from these samples to construct a model population. The model's parameters could be varied to ask what effects the routine work of gamekeepers might be having on the dynamics culled populations (McDonald & Harris, 1999).

The median ages at death caused by gamekeepers were 11·6 and 8·0 months for male and female stoats, and 9·3 and 9·2 months for male and female weasels. Gamekeepers on traditional estates trap mustelids all year round, but especially when they are most likely to threaten sitting birds and chicks. So just over 40% of the samples for both species were collected in spring, leading to a seasonal increase in mortality rate (McDonald & Harris, 2002).

Of 40 stoats sent for tooth cementum analyses (Sect. 5.1), the oldest was age 4; likewise, 2 of 38 weasels had reached age 2. Life tables estimated a first-year survival rate for male and female stoats as 0.41 and 0.26, and 0.15 and 0.03 for weasels. Model weasel populations continued to increase (λ = 1·35) despite culling; model stoat populations decreased slightly (λ = 0.95). Both were strongly affected by variation in food supplies, but the higher productivity rate of weasels made them remarkably tolerant of trapping.

Concentrated trapping in spring is especially likely to affect stoat numbers when lactating females are removed before their kits reach independence. Depleted local populations can soon be restored by immigration, because even large game estates are surrounded by un-keepered land, and long-distance dispersal by young stoats is common. But gamekeepering did not seem to be contributing to their disappearance.

The New Zealand context is totally different, because stoats are pests, and population models are being used to ask quite different questions. All of them emphasise the primary question of how to protect native fauna. The short but damaging post-seedfall irruptions of mice in New Zealand beech forests (Sect. 5.3) parallel the population dynamics of lemmings and voles in the stoats' native environments in the northern hemisphere, so the stoats brought to New Zealand were pre-adapted to respond to the masting cycle.

The New Zealand literature on the consequences of the beech masting cycle provides two well worked examples. The earlier models concentrated on the impact of peak-year trapping—that is, how many stoats have to be removed to *reduce the damage* that peak stoat populations can do to native fauna. Later models were able to specify the best time to *prevent a peak* population from developing in the first place. Because all these models were built from field data, they were descriptive models rather than predictive models that have been tested and shown to be able to predict demographic responses. The latter models might be called "preventive descriptive models".

In early 2000, several experienced population modellers had access to the right software running basic population models assuming a stationary population. But populations of stoats in New Zealand are seldom, if ever, stationary, especially in beech forests during a masting event (Chap. 5). A population model for beech forest stoats needs to incorporate unstable stoat population numbers, and to show the best time to take action before the next plague, and protect resources between plague years. That means regular field monitoring is needed to help plan the most effective management strategy. Is that extra cost worth the possible savings by culling only in high-risk years?

The mohua (*Mohoua ochrocephala*) is a small endemic passerine restricted to southern beech forest (Sect. 14.2.2), vulnerable to accelerated mortality during a masting year (Sect. 5.4). David Choquenot constructed a demographic model to ask whether the cost of trapping stoats to protect sitting female mohua in high-risk seasons (then estimated at \$7.55/ha/year) could be offset by not trapping when the risk was deemed to be low, depending on the cue employed. For example, a specified conservation objective for mohua (<10% probability of summer adult density falling below 0.2/ha) could be achieved most efficiently by undertaking stoat control only when cued by a threshold level of mouse abundance (1.5 mice/100 trap nights). Cueing stoat control according to stoat density reduced the frequency of control operations required to achieve the specified level of protection for mohua by 14%. The overall cost of this strategy, however, was appreciably higher than the alternative of cueing stoat control by mouse density, because the savings achieved through a reduced number of control operations were outweighed by the higher cost of stoat monitoring (Choquenot, 2006).

Another question probed by Nigel Barlow and Mandy Barron concerns whether fertility control or culling would be most effective in managing stoat populations (Barlow & Barron, 2005). These models, based on published means for stoat life history parameters (Chaps. 4 and 5), showed that not all standard trapping operations were equally effective in protecting threatened fauna. Trapping in podocarp forests was more effective than trapping elsewhere, mainly because populations of stoats living there had a lower rate of increase (Chap. 6). The models suggested that a hypothetical sterilisation campaign would yield results similar to those from trapping.

The problem was that the effects of trapping on the dynamic annual fluctuations of stoat populations through the masting cycle cannot be realistically represented by standard models that do not incorporate the different dynamics of stoat populations during different stages of the beech masting sequence. Without access to the age-specific and annual variations in fertility and natural mortality rates calculated from primary data, models cannot explore the impacts of management on post-seedfall stoat irruptions.

Effects of the Masting Cycle Given a large enough sample of stoats from known areas and times, and extensive data on age, survival and reproductive condition (Chap. 4), it became possible to calculate annual variations in age-specific fertility and mortality rates for three beech forest stoat populations (Chap. 5). Successive generations of stoats that followed an irruption year made different cohort-specific contributions to the variations on stoat numbers through the 3–5 years of an average masting sequence (Wittmer et al., 2007). The wild swings in cohort-specific productivity and mortality rates (Table 5.1) produced different age structures from 1 year to the next. So, estimates of mean age and kill rate depend on the natural mortality rates, which are never stable from year to year.

The finite rate of increase (lambda λ), calculated using a linked series of Leslie matrices representing the stages of the masting sequence, varied from strongly positive, $\lambda = 1.98$ during post-seedfall summers to negative, $\lambda = 0.58$ during non-seedfall summers. Mean population growth rates through the sequence swung from 0.85 (negative) to 1.13 (irrupting), mainly due to high reproductive success of adult females and high survival of young when mice were abundant, combined with low fertility and low survival of young females when mice were few (Fig. 5.3). Any realistic model of stoat population dynamics in beech forests has to include this sequence of within- and between-year variations, and there was only one data set of these available.

The model developed by Roger Powell used the key parameters provided by Wittmer's analysis to develop a 4-matrix model describing the effects of a single trap line sampling stoat populations through the beech cycle (King & Powell, 2011: fig. 2). The numbers of stoats culled at different intensities was set for a model population starting with 140 stoats in a 40 km^2 model study area, similar to the number of stoats living within the area of the Eglinton Valley. The model assumed that immigration continued throughout the trapping programme, and that immigration was greater when culling intensity was greater, or even matched it. It covered masting sequences both with and without culling, at levels of effort ranging up to the highest for which we had real data from the Eglinton Valley for the period 1976–78: 400 m trap spacing, traps baited and checked daily 14 days/month, totalling approximately 30,000 trapnights/year over 40 km^2.

For a trapping effort of 30,000 trapnights a year, the model predicted that 88 females and 96 males stoats would be removed, totalling 184, and the population left after immigration and culling would be 78 stoats.

The much more permanent, intensive trap line set up in the Eglinton Valley in late 1997 had 193 trapping stations spaced at 200 m intervals for c. 40 km, following the road along the length of the Eglinton valley, set permanently and checked monthly (Dilks et al., 2003). A forested saddle at the head of the valley allowed immigration by stoats across the Main Divide from the Hollyford Valley, where there was no stoat trapping. This more recent Eglinton operation was complicated by high numbers of ship rats, which also respond to a heavy seedfall.

Occupied or damaged traps unable to catch again until next month reduced the total potential of about 70,000 trapnights a year, especially during irruption years. The model calculated that the average total kill to be 235 stoats, of which >50% would be females, leaving 40 survivors after allowing for variation in productivity and immigration. Catches from 2001/2002–2007/2008 averaged 140 stoats (range 80–169) in non-seed years, compared with 308 in seed year 1999–2000, and 400 in 2000–2001.

Of course, traps set in a single line can achieve only a low intensity stoat suppression regime. Nevertheless, it permitted a significant increase in kākā breeding success (80% fledging) and juvenile survival (100%) by comparison with other beech forests with no protection from stoats (Sect. 14.2.7).

Apparent Versus True Culls An essential point shown by Roger's model is the difference between an apparent cull (the total number killed), equivalent to sustained harvesting, and a true cull, the number of stoats killed exceeding the natural mortality rate of undisturbed populations, which thereby achieves a real drop in density.

During the summer of a post-seedfall irruption, large numbers of stoats are killed, but nearly all are young of the year in poor condition, of which few would have lived to 1 year old. Consequently, trapping mortality is seldom additive above the high natural death rate of stoats for that year (Table 5.1). For every ten stoats killed during an irruption, nine would have died before their next birthday anyway, so few lived to achieve any reproductive success (Sect. 5.2).

By contrast, in non-seedfall years, fewer stoats are born but most are in relatively good condition and their survival over the winter is near 0.5 for all ages. Natural mortality is low, so more of those killed would otherwise have been expected to live to breed. Consequently, for every ten stoats killed then, five might otherwise have survived, implying a possible 50% reduction in number of stoats available to breed in the following season. This is true culling, which can have a real effect on density (King & Powell, 2011).

The essential conclusion is that the best time to reduce stoat damage to wildlife is to target females when their numbers are still small, but food supplies are good and rapidly increasing (Sect. 4.5). Their potential fertility is high, so removing them *before* they produce a peak year cohort is the best time to undertake a landscape-scale aerial operation. But that is also the time when only a few stoats are present and they are doing little obvious damage, so the incentive and funding to launch a control operation are minimal.

Expensive trapping programmes are certainly vitally necessary for after-the-fact mitigation of *damage* to nesting birds during an irruption, but they are too late and too small to have any more than minimal effect on long term control of stoat *populations*. Moreover, if a target population includes a large proportion of innately trap shy stoats, estimates of trapping success could be quite wrong.

On the other hand, even a low level of trapping effort sustained year round can achieve real increase in fledging success for the bird species most vulnerable to stoats (Chap. 14). That is still worth doing, even though eradication will be possible only when we develop the technology to turn off the tap bringing in replacements for the stoats killed.

11.5 Do Stoats Control Their Prey, or Vice Versa?

What effect does predation by stoats have on their prey, especially if those prey are also pests in their own right?. If damaging stoat irruptions are primarily driven by a sudden increase in rodent numbers, trapping the few stoats present during the early stages of the sequence needs to be intensive to prevent it, but it never is. If peak numbers of stoats decline as rodents disappear in both trapped and untrapped areas, their loss cannot be credited to successful trapping. We need more information about years with both low and high stoats numbers, because the two scenarios have important implications for predator management in masting forests.

Passengers, Not Drivers One well-documented assemblage of rodents and mustelids observed over 750 ha of North Island forest offered an opportunity to test four ideas going back to decades of work on lemming cycles, but which also apply here. The main question to be settled is whether predators living through a rodent irruption are 'Drivers or Passengers' (Blackwell et al., 2003).

The four conclusions were as follows:

1. Predators can slow but not prevent a post-seedfall irruption of rodents. Populations of mice and rats irrupted to high densities in areas both with and without suppression of stoats.

2. Predators cannot reduce peak rodent population densities. Peak densities of mice and rats were not significantly different between treatment and non-treatment areas.
3. Predators can only marginally hasten the rate of decrease in rodent populations during the crash phase. Numbers of mice and rats did not decrease significantly more slowly where stoats were removed.
4. Predators can sometimes limit low-phase prey populations, depending on forest type. Populations of rats in beech forest, and population of mice and rats in coastline habitats, were significantly higher in areas with predators removed, but in tawa-podocarp forest predator removal had no effect.

This study confirmed previous observations that sudden pulses in beech seed abundances drive the unstoppable early stages of post-seedfall rodent and stoat irruptions. The effects of predation on rodents during the decrease and low phases is unclear.

A simple way to judge the truth of a statement, in science as in life generally, is to ask if it is consistent with everything else we know about the subject. So if these conclusions are correct, how do they affect so-called 'common knowledge' about what stoats are doing in New Zealand forests? It seems that real stoats do not match their reputations as voracious predators—on the contrary, they are more vulnerable to shortages of food than is usually assumed. Predation by stoats is not heavy enough to affect post-seedfall rodent irruptions—rather, stoats are simply carried along for the ride. So early models of predator-prey interactions making simple assumptions of stoat kill rates on rodents might be quite wrong.

Stoat Impacts on Rodents The critical parameter, the kill rate, depends on the functional response (FR) of a predator to increasing densities of prey (Jones et al., 2011). The functional response of stoats can be estimated from diet reports if combined with estimates of energy requirements and prey availability. First, existing field indices of stoat abundance, as indexed by tracking tunnels, converted into indices of stoat abundances and estimates of minimum kill rates from stoat gut contents, and supplemented by observations of feeding activity and estimates of gut-passage time from captive animals were assembled. The models were constructed to fit the seasonal variation in numbers of mice eaten against mouse density from Fiordland data reported in Chap. 5, and the same for rats from the Department of Conservation's data on rats eaten at Okarito (Murphy et al., 2008).

Male stoats need to find some 90 g of meat a day to support their galloping metabolism, females somewhat less except when feeding kits. New Zealand beech forests are marginal habitats for them, especially during a post-seedfall crash in mouse numbers. Stoats are indeed efficient specialist predators of mice, but mice are small and not common. When living only on mice, a male stoat would need to catch 4–6 mice at 15–20 g each per day to get their 90 g. But the model calculations based on field data suggested that, in autumn and winter, stoats killed a theoretical maximum of only 1.13 mice per stoat per day. In spring and summer, the maximum kill rate was calculated to be 1.04 mice per stoat per day. The mean density of mice in beech forests was estimated by Innes et al. (2010) as 4.4/ha for mice. During a brief post seedfall summer peak, mouse density might reach 50/ha, but very rarely and not for long.

In practice, very few real density estimates for mice are available, only indices of abundance, which are not the same thing, and most of them suggest populations far too low to support stoats eating nothing else (Murphy & Nathan, 2021). No wonder stoats also need to hunt birds and invertebrates as well. Even then, it looks as if stoats have to work hard to hunt down every last mouse. Only in good years can their reward can be worth the effort, because then even a modest mouse peak can hugely boost a stoat's chances of reproductive success (Sects. 4.5 and 5.3).

The data available on rats were more limited, but suggested a maximum of 0.15 rats killed per stoat per day from a population of about 3.9/ha. A single ship rat weighing 120–160 g could meet a male stoat's daily food requirement all at once, minimising the need for it to spend energy on further hunting that day. It looks as if previous models have overestimated the stoats' usual rate of killing rodents in forests. If true, that conclusion has implications for ideas about whether effective suppression of stoats in some forests but not others can increase the risk of mesopredator release of rats (Whitau et al., 2023).

Kiwi-Hunting Specialists? All descriptive models depend on acquiring data from somewhere, preferably from the field. Nonetheless, theoretical questions can be investigated using imaginary numbers that span realistic ranges relevant to important concerns. One such question addressed the effects of different dietary preferences among stoats (McAulay et al., 2021). For example, one individual stoat with a taste for kiwi could do a lot more damage than a hypothetical 'average' stoat. Garcia-Diaz et al. (2021) presented the output of a model of a kiwi population living within range of a local stoat population that included different proportions of specialist and generalist kiwi hunters. Their conclusion was that the size of a stoat's home range was more important than the density of kiwi specialists. The larger an individual's home range, the more likely it would be to encounter a kiwi. Home ranges expand when prey are scarce (Sect. 9.3), which only adds to the risk of sudden death for kiwi chicks emerging from their burrows during a season when rodents are few.

11.6 Risk of Mesopredator Release

All models depend on multiple assumptions, that, if false, can produce cascading errors in the results. But models can also generate hypotheses that can be tested in the field. One important question that has worried conservationists for years is whether a successful stoat eradication campaign might allow a surge in numbers of rats. Stoats are important members of integrated animal communities with complex links in all directions. Therefore, the naïve assumption of early management programmes (Sect. 1.2), that one member of a community can be removed without affecting others, was often wrong. On the contrary, targeted pest control can easily generate unexpected consequences.

Rats Animal communities are structured, with an apex predator at the top that restricts the numbers and activities of less dominant predators further down the hierarchy. In forests, removal of a top predator like the stoat or cat could theoretically release the lower-level species, ship rats and mice. These subordinate species can reach much higher numbers than can stoats, so can cause more damage. Wendy Ruscoe and team set up a large, replicated Before-After Control-Impact field experiment to ask whether this ever happens in the wild (Ruscoe et al., 2011).

In each of four study locations of about 900 ha each in mixed podocarp—broadleaved forest in the North Island, two study sites 2 km apart were set up for separate treatment versus non-treatment observations. All target species belonging to a four-species assemblage were first monitored whilst undisturbed. Then, in separate experiments, members of each species were removed using methods having the least possible effects on the rest of the local community.

There was no measurable response by rat and mouse populations to removal of the assumed top predator alone (stoats), but there was a strong competitive release of rat numbers following removal of possums alone, and strong behavioural response of mice following removal of rats alone. Hence, there were indeed two important mesopredator release effects, but in the podocarp forest communities observed, neither involved stoats.

Because replicated, randomised field experiments are seen as the gold standard of scientific knowledge, especially when they are consistent with much else we know about the species individually, this result settled the argument for quite a while. Moreover it seemed to confirm the calculations of Jones et al. (2011), that models tend to overestimate the effects of stoat predation on forest rodents , and to confirm Blackwell's (2003) description of stoats as passengers, not drivers, of post-seedfall rodent irruptions. Worries about the risk of mesopredator release of rats after effective stoat control faded into the background. Until a new study cast doubts on that conclusion.

The Rotoiti Nature Recovery Project, in the beech forests of Nelson Lakes National Park, has been managed by the Department of Conservation since 1997. For this research, four treatment sites and an adjacent non-treatment site in the same area were used, plus another two comparable treatment sites in Kahurangi National Park. All relevant records of predator control and tracking tunnel operations on the six treatment sites were gathered and tested for the responses of ship rat populations to variables including a stoat culling treatment, possum culling, rat culling, elevation, beech seedfall and previous rat population indices across all sites, and the results analysed by Kelly Whitau (Whitau et al., 2023). The total database listed 6690 tracking tunnel records for the years 1998–2016. This period included three particularly high seedfall years in 2000, 2006, and 2014, all exceeding the threshold of 2000 seeds m^2 that triggers a likely rodent outbreak.

The model confirmed some trends already known, such as the general scarcity of rats at high elevations and their population increases after mast years. But, contrary to previous studies, it also showed that reduced numbers of stoats were associated with clearly increased ship rat populations. Why the contradiction?

The relative numbers of rats, mice and stoats vary widely across different habitat types, and the result of removing one predator more effectively than another may depend on the starting populations of each. This pattern was already suggested by comparing two early long-term surveys in different forest types using the same fixed protocols. In the 1970s, intensive trapping in beech forest of northern Fiordland caught variable numbers of mice, stoats and no rats, whereas in the 1980s the same method applied in North Island podocarp forest at Pureora found huge numbers of rats and few mice or stoats (Chaps. 5 and 6). Well-designed research is needed to explore the difference.

Another team led by Susan Walker used 18 years of 3-monthly records from a largescale tracking-tunnel network to define six classes of forest rodent population dynamics (Walker et al., 2019). The classes ranged from 'irruptive' (in cool beech forests where mice and stoats are generally scarce but can occasionally surge to great numbers, and rats are few) to 'continuously ratty' in warmer podocarp forests where mice and stoats are few and rats are abundant all the time.

The results summarised the differences among forests representing the spectrum of conditions. In cool forests, low median but synchronous peaks in rat and mouse tracking rates after resource pulses are likely, especially in mice alone where rats are absent (Chap. 5). Low temperature is a stronger limitation for ship rats than for mice, perhaps because ship rats were originally a tropical species. At the other extreme, rats can reach high numbers at the expense of mice in warm forests, so reciprocal distribution of high rat and low mouse tracking rates are more likely (Bridgman et al., 2018).

A follow-up analysis compiled a different dataset including ship rat, mouse and mustelid footprint tracking data collected from across New Zealand's public conservation lands between April 1996 and June 2022. It was already clear from previous work how indices of rodent activity had varied with habitat and temperature across the landscape. The question of most concern now was: can conservation managers predict the activities of stoats from those patterns of rodent indices?

Rodent and stoat activity patterns were indeed linked, but on a broad scale, the highest abundance indices for rodents and stoats did not always match. Mice drove short-lived post-seedfall irruptions of stoats in cool eastern beech forests, but in western beech forests, mice and rats irrupted together, and ship rats remained available for longer. In the warm podocarp-broadleaved forests of the North Island, abundant rats provided rich resources for stoats, and yet stoat tracking rates there were low.

The team concluded that stoats are stuck between two more or less equally challenging alternatives. In the cool beech forests where the masting cycle orchestrated landscape-scale swings between feast or famine, stoats had priority access to peak numbers of rodents, but had to survive the poor years between mast years. In the warm, rich environments where rats were abundant every year, stoats had to compete with cats and ferrets for rats, whilst also avoiding becoming prey to those larger predators. In short, the team concluded, stoat numbers may be limited by food availability in cool environments, but by top-down suppression by feral cats or ferrets in warm and dry environments (Walker et al., Unpublished).

So what about the apparent contradiction in studies of mesopredator release? There is no contradiction, because, as was already known, animals usually behave differently in different habitats under different climatic conditions. On the one hand, mesopredators were not released by removing stoats from ratty, North Island habitats where stoats are normally few compared with other predators (Ruscoe et al., 2011). On the other hand, in the South Island forests where rats breed much faster than stoats, and other predators were few, removing stoats can increase the chances of precipitating a classic mesopredator release in rats (Whitau et al., 2023). Understanding stoat distribution and behaviour depends on habitat, more than we ever thought we knew from simple observation and untested descriptive models.

References

Alterio, N., Moller, H., & Brown, K. (1999). Trappability and densities of stoats (*Mustela erminea*) and ship rats (*Rattus rattus*) in a South Island *Nothofagus* forest, New Zealand. *New Zealand Journal of Ecology, 23*, 95–100.

Barlow, N. D., & Barron, M. C. (2005). Modelling the dynamics and control of stoats in New Zealand forests. *Science for Conservation, 252*, 1–40.

Bartolucci, F., & Pennoni, F. (2007). A class of latent Markov models for capture–recapture data allowing for time, heterogeneity, and behavior effects. *Biometrics, 63*, 568–578.

Blackwell, G. L., Potter, M. A., McLennan, J. A., & Minot, E. O. (2003). The role of predators in ship rat and house mouse population eruptions: Drivers or passengers? *Oikos, 100*(3), 601–613. https://doi.org/10.1034/j.1600-0706.2003.11026.x

Breedt, B., & King, C. M. (2021). Distribution and detectability of mammalian pests in the Waikato region. *New Zealand Journal of Zoology, 49*, 37–52.

Bridgman, L., Innes, J., Gillies, C., Fitzgerald, N., Rohan, M., & King, C. M. (2018). Interactions between ship rats and house mice at Pureora Forest Park. *New Zealand Journal of Zoology, 45*, 238–256. https://doi.org/10.1080/03014223.2018.1464477

Choquenot, D. (2006). Bioeconomic modelling in conservation pest management: Effect of stoat control on extinction risk of an indigenous New Zealand passerine, Mohua ochrocephala. *Conservation Biology, 20*(2), 480–489.

Christie, J. E., Brown, D. J., Westbrooke, I., & Murphy, E. C. (2009). Environmental predictors of stoat (*Mustela erminea*) and ship rat (*Rattus rattus*) capture success. *Department of Conservation Research and Development Series, 305*, 1–27.

Dilks, P. J., & Lawrence, B. (2000). The use of poison eggs for the control of stoats. *New Zealand Journal of Zoology, 27*, 173–182.

Dilks, P., Willans, M., Pryde, M., & Fraser, I. (2003). Large scale stoat control to protect mohua (*Mohoua ochrocephala*) and kaka (*Nestor meridionalis*) in the Eglinton Valley, Fiordland, New Zealand. *New Zealand Journal of Ecology, 27*, 1–9.

Efford, M. G., & Fewster, R. M. (2012). Estimating population size by spatially explicit capture–recapture. *Oikos, 122*, 918–928.

Garcia-Diaz, P., Binny, R., & Anderson, D. P. (2021). How important is individual foraging specialisation in invasive predators for native-prey population viability? *Oecologia*. https://link.springer.com/article/10.1007/s00442-020-04814-6

Glen, A. (2014). Animal pests: Residual trap catch index for possums. In T. Greene & K. McNutt (Eds.), *Biodiversity inventory and monitoring toolbox*. Department of Conservation.

Innes, J., Kelly, D., Overton, J. M., & Gillies, C. (2010). Predation and other factors currently limiting New Zealand forest birds. *New Zealand Journal of Ecology, 34*(1), 86–114.

Johnstone, K. C., & Garvey, P. M. (2023). To boldly go: Methods to quantify personality in mustelids. *Animal Behaviour, 202*, 139–147.

Johnstone, K. C., Price, C., & Garvey, P. M. (2024). Personality, sex and capture biases: Challenges for predator monitoring and management. *Journal of Applied Ecology*. https://doi.org/10.1111/1365-2664.14724

Jones, C., Pech, R., Forrester, G., King, C. M., & Murphy, E. C. (2011). Functional responses of an invasive top predator *Mustela erminea* to invasive meso-predators *Rattus rattus* and *Mus musculus*, in New Zealand forests. *Wildlife Research, 38*(2), 131–140. https://doi.org/10.1071/wr10137

King, C. M., & McMillan, C. D. (1982). Population structure and dispersal of peak-year cohorts of stoats (*Mustela erminea*) in two New Zealand forests, with especial reference to control. *New Zealand Journal of Ecology, 5*, 59–66.

King, C. M., & Powell, R. A. (2011). Managing an invasive predator pre-adapted to a pulsed resource: A model of stoat (*Mustela erminea*) irruptions in New Zealand beech forests. *Biological Invasions, 13*, 3039–3055.

King, C. M., & White, P. C. L. (2004). Decline in capture rate of stoats at high mouse densities in New Zealand *Nothofagus* forests. *New Zealand Journal of Ecology, 28*, 251–258.

King, C. M., Davis, S. A., Purdey, D. C., & Lawrence, B. (2003). Capture probability and heterogeneity of trap response in stoats, *Mustela erminea*. *Wildlife Research, 30*, 611–619.

King, C. M., McDonald, R. M., Martin, R. D., MacKenzie, D. I., Tempero, G. W., & Holmes, S. J. (2007). Continuous monitoring of predator control operations at landscape scale. *Ecological Management and Restoration, 8*, 133–139.

King, C. M., Innes, J. G., Gleeson, D., Fitzgerald, N., Winstanley, T., O'Brien, B., Bridgman, L., & Cox, N. (2011). Reinvasion by ship rats (*Rattus rattus*) of forest fragments after eradication. *Biological Invasions, 13*, 2391–2408.

McAulay, J. R., Monks, J. M., Wilson, D. J., & Seddon, P. J. (2021). Individual specialists within a generalist niche: Variable diet of stoats and implications for conservation. *New Zealand Journal of Ecology, 45*(2), 3443.

McDonald, R. A., & Harris, S. (1999). The use of trapping records to monitor populations of stoats *Mustela erminea* and weasels *M. nivalis*: The importance of trapping effort. *Journal of Applied Ecology, 36*(5), 679–688.

McDonald, R. A., & Harris, S. (2002). Population biology of stoats *Mustela erminea* and weasels *Mustela nivalis* on game estates in Great Britain. *Journal of Applied Ecology, 39*, 793–805.

Murphy, E. C., & Nathan, H. W. (2021). House mouse. In C. M. King & D. M. Forsyth (Eds.), *The handbook of New Zealand mammals* (3rd ed., pp. 225–242). CSIRO Publishing.

Murphy, E., Maddigan, F., Edwards, B., & Clapperton, K. (2008). Diet of stoats at Okarito Kiwi Sanctuary, South Westland, New Zealand. *New Zealand Journal of Ecology, 32*(1), 41–45.

Ruscoe, W. A., Ramsey, D. S. L., Pech, R. P., Sweetapple, P. J., Yockney, I., Barron, M. C., Perry, M., Nugent, G., Carran, R., Warne, R., Brausch, C., & Duncan, R. P. (2011). Unexpected consequences of control: Competitive vs. predator release in a four-species assemblage of invasive mammals. *Ecology Letters, 14*(10), 1035–1042. https://doi.org/10.1111/j.1461-0248.2011.01673.x

Vattiato, G., Binny, R. N., Davidson, S. J., Byrom, A. E., Anderson, D. P., Plank, M. J., Carpenter, J. K., & Alex James, A. (2023). Detectability of ten invasive mammal pests in New Zealand: A synthesis of spatial detection parameters. *New Zealand Journal of Ecology, 47*, 3552.

Walker, S., Kemp, J. R., Elliott, G. P., Mosen, C. C., & Innes, J. G. (2019). Spatial patterns and drivers of invasive rodent dynamics in New Zealand forests. *Biological Invasions, 21*, 1627–1642.

Walker, S., Kemp, J. R., & Elliott, G. P. (Unpublished). *Temporal and spatial patterns of rodent and mustelid tracking rates across New Zealand public land between 2005 and 2022.*

Watkins, A. F., McWhirter, J. L., & King, C. M. (2010a). Patchiness in distribution of rodents and mustelids in New Zealand forests. *Wildlife Research, 37*, 13–18.

Watkins, A. F., McWhirter, J. L., & King, C. M. (2010b). Variable detectability in long-term population surveys of small mammals. *European Journal of Wildlife Research, 56*, 261–274.

Whitau, K., Kelly, D., Galloway, T. N. H., MacFarlane, A. E. T., van Vianen, J. C. C. M., Rossignaud, L., & Doherty, K. J. (2023). Effects of altitude, seedfall and control operations on rat abundance in South Island Nothofagus forests 1998–2016. *New Zealand Journal of Ecology, 47*, 3502.

Wittmer, H. U., Powell, R. A., & King, C. M. (2007). Understanding contributions of cohort effects to growth rates of fluctuating populations. *Journal of Animal Ecology, 76*(5), 946–956. https://doi.org/10.1111/j.1365-2656.2007.01274.x

Genetics 12

Abstract

Genetic technology is a transformative new tool now essential for stoat management. It can distinguish survivors from reinvaders after an apparently successful eradication. On Secretary Island in 2015, 205 stoats caught on the island and 40 from the adjacent mainland confirmed a resident population of trap-shy females on the island, supplemented by occasional swimmers across Acheron Passage. After a successful eradication operation on Rangitoto, a single stoat found on the island was identified as reinvader from the mainland, 3 km away. Stoats survive because their high productivity normally compensates for high natural or imposed mortality, but early experiments in chemo sterilisation were not encouraging, and depended on edible baits that could be refused. By contrast, gene drives work by changing the natural patterns of inheritance to produce only young of one sex, so the population should eventually die out. But the development of a heritable gene drive would be impossible for stoats with a generation time of 2 years, and modelling of possible outcomes predicted uncertain results and the rapid appearance of mutations conferring resistance. New Zealand's history of releasing mustelids as irretrievable self-perpetuating pest control agents is a stark warning against the modern equivalent of the same logic.

Keywords

DNA · Reinvaders versus survivors · Chemo sterilisation · Gene drives · Social licence

C. M. King, *Stoat in the Dock*, Fascinating Life Sciences,
https://doi.org/10.1007/978-981-96-4562-6_12

12.1 Genome Sequencing

The DNA of all mammals carries a detailed record of their ancestry and distribution going back for generations. That genetic information can now be read nearly as easily as any literate person can read a book, but it takes the right skills and equipment, and these have become available on the scale required only very recently. The impact of genetic analyses on our understanding of the natural world, especially the business of stoat control, is proving as significant as the impact of the printing press was on medieval scholarship, and for similar reasons.

12.2 Case Examples

Population genetic analyses can identify the origins of a pest incursion, and after a control programme, distinguish between survivors and reinvaders. The first geneticist to develop this powerful technique for stoats, Andrew Veale, used it to ask why the eradication programme on Secretary Island, which started in July 2005, had not yet succeeded (Sect. 7.3). He genotyped 205 stoats caught on the island, plus 40 stoats from the opposing mainland coast, and determined the age and sex for each one. His results confirmed the survival on the island of a core group of trap-shy adult females which continued to breed new generations offspring, most of which were soon removed by trapping. The mothers remained on separate territories mapped across the island. Infrequent but unstoppable immigration from the mainland across the narrow Acheron Passage, an easy swim for a stoat (Sect. 7.2) continues, especially after a beechmast year (Veale et al., 2013).

The majority of recruitment comes from resident stoats breeding, at a rate probably capable of maintaining the population indefinitely under the current management regime. The question remains of whether, if the resident population was removed and trapping was ongoing, could immigration maintain a new population on the island?

In 2009, Rangitoto/Motutapu Islands in the Hauraki Gulf were cleared of all mammalian pests. A year after this eradication was completed, a single stoat was trapped on the island (Sect. 4.2). Genetic population assignment deduced that this individual was a reinvader, which had probably swum from the adjacent mainland via the Tamaki Estuary, 3 km away (Veale et al., 2012). This swimming distance seemed extraordinary at the time, far greater than any previously known stoat incursions, and would not have been believed except that this individual's genetic fingerprint was firmly liked to the mainland population. It led directly to the experiments described in Sect. 7.2.

Because New Zealand leads the world in per capita funding for research on mustelids, it seems fitting that we have contributed substantially to international studies on their genomics. A male stoat trapped in the Waikato Region is now the 'genomic type specimen' for stoats. In a project led by Andrew Veale and funded by the Bioheritage Science Challenge and Predator Free 2050, in collaboration with the international Vertebrate Genome Project, this stoat's genome was sequenced with a range of different technologies and assembled to give a gapless, error-free complete assembly of the genome. This is now one of the highest quality genomes of any species in the world, and it will provide the basis for future research that may assist in developing novel control tools.

Similarly, a male weasel trapped outside Brook Waimarama Sanctuary near Nelson has just had a similar quality genome assembled, and this specimen will be the genomic type specimen for weasels. These genomes are already being used to great effect in research. In Andrews PhD research study, he used up to 17 genetic markers to examine the connectivity of stoat populations. Now, in his ongoing (currently unpublished) research, he is using over 60,000 markers to obtain extreme fine-scale relatedness and at times full pedigrees (Veale, 2024).

12.3 Gene Drives

One feature that stoats and all other small invasive predators have in common is very high natural mortality combined with very high fertility. Their population numbers depend on rapid replacement of individuals lost, which is normally easy because in most years many more young are produced than are needed to replace lost adults. High natural mortality also means that human efforts to control stoats by lethal methods has to remove more individuals than will die anyway, and that additional toll on their numbers is very hard to achieve. On the other hand, the requirement for high breeding success year after year makes such populations vulnerable to bad breeding seasons, and also to any method of reducing fertility that would minimise the number of stoats that have to be killed (Hinds et al., 2000).

Only now (2025) is research on fertility control moving away from old ideas of using baits and chemo sterilants into potential new solutions based on genetics. The huge funding required, and the public debate over the control methods needed, are issues to be resolved at the level of social and human decisions, rather than that of stoat biology.

If reducing fertility is so important, could we edit the reproductive instructions of stoats in ways that would alter the normal laws of inheritance? For example, their sex ratio at birth is nearly 1:1, even though only a few dominant males get to breed (Sect. 4.4.1). But no breed-fast-die-young species can survive for long if more than a few generations could be made to produce offspring of only one sex. Since nearly all females breed, but not all males, the key is to reduce the reproductive success of females.

The problem is that the whole apparatus of reproduction has evolved to defeat any such risk. Natural birth rates of pests are managed by genes selected over uncountable generations for their ability to maximise the potential number of surviving offspring produced in any conditions.

Gene drives work by changing the natural patterns of inheritance (Dearden et al., 2017). With enough modified individuals introduced into a wild population, and enough generations of daughterless offspring, the population should—in theory—eventually die out. Not a single individual would need to meet death by any of the currently unavoidable standard methods which make animal welfare activists shudder. The effect could spread across the country at landscape scale, into all the remote places where any human action is impossible.

This sounds, at first, like an ideal answer to those who have long doubted the feasibility of the Predator Free New Zealand 2050 programme (Chap. 13). But in fact, this technique is less like an offer of national salvation from intractable pests, and more like a modern version of the pastoralists' tragic mustelid introduction programme (Sect. 1.2). Rabbits in the nineteenth century had resisted all known methods of control; runholders sought a self-perpetuating rabbit-killing mechanism; they believed mustelids would supply it; they didn't, and now the failed alien predators cannot be recalled, while rabbits still thrive. Mustelids in modern New Zealand continue to resist all known methods of control; targeted gene drives offer a self-perpetuating pest-removing mechanism; if they don't, then in the future we will be left with engineered genes in nature that cannot be recalled, while the mustelids develop resistance and thrive.

Modelling of potential outcomes has found that the chances of eradicating any pest population with a gene drive are far from certain, and mutations conferring resistance are to be expected. Social, legal and political problems are likely to be even more difficult to overcome (Mercier et al., 2024), especially on the international stage (Espeut & Gemmell, 2017). Worse, gene drives work on generational timescales, leading to a population collapse generally predicted (in modelled situations) to take 30+ generations. Recent modelling suggests that the outcome of a proposed landscape scale operation will be influenced by demography and life-history traits (Birand et al., 2022). For rodents with a generation time of 6 weeks, the potential for eradication is an attractive prospect, but not for stoats with a minimum generation time of 2 years (Sect. 4.5).

Thoughtful people at the centre of the debate agree that the benefits of this sort of gene technology are alluring, especially in the fields of medicine and disease control. But the results could so easily turn out to be the opposite of those expected (Sect. 1.2), and if it did, the national and international consequences would be incalculable.

Of all countries in the world, New Zealand is the one best placed to learn from its own history. We know all too well what happens when policies that seem quite reasonable turn out to have drastic unforseeable effects. Especially, we have learned the hard way (e.g., Sect. 5.1) why it is very important that that debate must be in the open, and that social licence has to be negotiated long before deployment. That means full public participation, especially with Māori, plus comprehensive international consultation. Beating out a broad social agreement on the potential use of gene drives will certainly cause a bigger row than the current repeated uproars over 1080 (Chap. 8).

That unhappy outcome is certainly to be avoided if possible. Macdonald et al. (2020) suggest that public engagement that acknowledges the range of underlying beliefs held by those for and against this new technology would help create a constructive dialogue on this thorny issue.

While the technological challenges of creating controllable gene drives are vast, those governing the social licence to use such technologies are worse. Gene editing technology commonly meets public resistance. History often reminds us how easily the complex web of ecosystem interactions can produce unexpected responses to even a simple management operation, especially if its supporters deny the chances of unwelcome consequences. Professor Newton's warning to the Colonial Government in 1876, about the folly of importing stoats and weasels was correct, timely and well argued, but ignored (Sect. 1.2).

In the context of this book, we need only emphasise, once again, that future debates on any self-perpetuating pest control mechanism need to include lessons from our own history of false beliefs and irretrievable introductions. That warning is becoming especially relevant now.

References

Birand, A., Cassey, P., Ross, J. V., Thomas, P. Q., & Prowse, T. A. A. (2022). Scalability of genetic biocontrols for eradicating invasive alien mammals. *NeoBiota, 74*, 93–103.

Dearden, P. K., Gemmell, N. J., Mercier, O. R., Lester, P. J., Scott, M. J., Newcomb, R. D., Buckley, T. R., Jacobs, J. M. E., Goldson, S. G., & Penman, D. R. (2017). The potential for the use of gene drives for pest control in New Zealand: A perspective. *Journal of the Royal Society of New Zealand, 48*, 225–244. https://doi.org/10.1080/03036758.2017.1385030

Espeut, K. M., & Gemmell, N. J. (2017). Conservation demands safe gene drive. *PLoS Biology, 15*, e2003850.

Hinds, L. A., Williams, C. K., Pech, R. P., Spratt, D. M., Robinsonm, A. J., & Reubel, G. H. (2000). Feasibility of immunocontraception for managing stoats in New Zealand. *Science for Conservation, 158*, 109.

Macdonald, E. A., Edwards, E., Balanovic, J., & Medvecky, F. (2020). Underlying beliefs linked to public opinion about gene drive and pest-specific toxin for pest control. *Wildlife Research*. https://doi.org/10.1071/WR19149

Mercier, O. R., Jones, N., King Hunt, A., Hemmerling, L., MacDonald, L., Horowitz, J., Kanli, R., Palmer, S., Belcher, S., & Lester, P. J. (2024). Views of conservation volunteers and environmental specialists on genetic technologies for pest control in Aotearoa New Zealand. *Frontiers in Conservation Science, 5*. https://doi.org/10.3389/fcosc.2024.1389930

Veale, A. J. (2024). Personal communication.

Veale, A. J., Clout, M. N., & Gleeson, D. M. (2012). Genetic population assignment reveals a long-distance incursion to an island by a stoat (*Mustela erminea*). *Biological Invasions, 14*(3), 735–742. https://doi.org/10.1007/s10530-011-0113-9

Veale, A., Edge, K. A., McMurtrie, P., Fewster, R. M., Clout, M. N., & Gleeson, D. (2013). Using genetic techniques to quantify reinvasion, survival and in situ breeding rates during control operations. *Molecular Ecology, 22*, 5071–5083. https://doi.org/10.1111/mec.12453

13 Predator Free New Zealand 2050

Abstract

Everything we know about stoats shows that they are remarkably resistant to eradication. Any new tool effective enough to exterminate them at national scale would be expensive and difficult to develop. Official attempts to fund the research needed started in 1999 with a grant to the Department of Conservation of NZ$6.6 million. It produced a new humane trap but no advances in fertility control. PFNZ2050 was launched in 2016, a hugely expensive programme with a much larger budget and a stated vision of restoring the dawn chorus that galvanised hundreds of community groups. They are now achieving impressive results at the local and temporary level, by routine and widespread deployment of conventional technology. Their work is essential to protect what biodiversity we have left until we can think of better ways to save it. Killing stoats is not as difficult now as preventing them from reinvading. Offshore islands are safe if >5 km from shore; the best mainland equivalents are fenced sanctuaries. By 2019 there were 14 of these, half ring fences, half on peninsulas. National implementation of these would cost about the same as PFNZ2050 and provide the best permanent protection for deeply endemic species available to date. Fence damage and unexpected incursions are possible but uncommon, and monitored 24/7.

Keywords

Official trapping campaigns · Local pest eradication · Community engagement in conservation · Fenced sanctuaries

Under New Zealand law it has been illegal to kill or possess any of the threatened native birds since the Wildlife Act 1953, but that could not help them while their native habitats (especially on the lowlands) were still being destroyed. Some 75% of New Zealand's original native vegetation has gone since human settlement. A

C. M. King, *Stoat in the Dock*, Fascinating Life Sciences,
https://doi.org/10.1007/978-981-96-4562-6_13

bird needs to find enough food to stay alive every day, but while it is not guaranteed to meet a predator that often, it is at daily risk of starving to death if it doesn't. The remains of a bird that died quietly for lack of resources are less often observed than the sight of a predator trashing a nest. So, although those of New Zealand's endemic birds most long adapted to the K-selected life style (low fertility and mortality, long adult life spans) are acutely vulnerable to accelerated predation, and opportunistic species are somewhat less so, all are vulnerable to the loss of their habitat and food supply. Hence, habitat protection and predator control are both absolutely necessary, but neither is by itself sufficient. The inspiring results of island eradications (Chap. 7) has generated a strong public desire to see similar results achieved on the mainland. As Victor Hugo, the famous French author, is supposed to have said, "Nothing is more powerful than an idea whose time has come".

13.1 The Origins of PFNZ2050

The problem with all current methods of predator control is that rats are too numerous and too rapidly replaced, and stoats are too widespread and too intelligent, for either species to be eradicated by trapping. Toxins are available and work on both better than anything else, but not in all circumstances. Ultimately, finding better ways to kill invasive predators is not the main issue anyway: the real problem is that reinvasion is impossible to prevent at any large scale, except on distant islands or behind fences. So in 1999 the New Zealand Government granted NZ $6.6 million to the Department of Conservation (DOC) to instigate a five-year program to find more cost-effective and sustainable approaches to managing stoats (Murphy & Fechney, 2003).

The new funding stimulated an explosion of research by various agencies, producing a new humane trap, the DOC 200/250, a review of international literature on diseases (McDonald & Larivière, 2001) and a discussion of the theoretical prospects of fertility control (Norbury, 2000). It also established New Zealand's first official research programme for captive management of stoats for fertility experiments, but concluded by confirming that stoat reproductive processes are far more difficult to disrupt than expected (Hinds et al., 2000). Meanwhile, the need for more and better methods of predator control in the wild rumbled on among conservationists, and gradually developed into a roar.

A Bold Idea In 2016, the New Zealand government initiated an ambitious national programme, called Predator Free New Zealand 2050 (PFNZ2050), allocating serious money (NZ$7 M/year) to seed-fund a national campaign to eradicate by 2050 the principal vertebrate pests perceived to be causing the ongoing collapse of our indigenous bird populations (Russell et al., 2015). The species targeted were, of course, the three mustelids (*M. erminea, M. nivalis* and *M. furo*), along with Norway rats and ship rats (*Rattus norvegicus and R. rattus*), and the Australian brushtail possum (*Trichosurus vulpecula*). The two European rats had stowed away with European shipping in the seventeenth and eighteenth centuries (King, 2019); mustelids and possums had been deliberately introduced (Sect. 1.2). The Polynesian rat

Rattus exulans is the smallest of the invasive rats, and by far the longest resident since it arrived in New Zealand with the ancestors of the Māori in the late 1200s, but it is now rare and retains important cultural values. Cats, mice and hedgehogs are excluded from the plan. Nobody targets them all anywhere; in practice, most operations focus on the worst two invaders, ship rats and stoats.

The idea of PFNZ250 was famously compared to President Kennedy's aim to put a man on the moon by 1969, a similar extravagant and visionary proposition also accompanied by powerful nationalistic optimism. In neither case was the necessary technology available at the time. PFNZ2050's initial reception was sometimes one of frank disbelief: one of my long-time correspondents in Australia emailed me to ask if it was a hoax. Even the prestigious journal *Nature* ran an article calling it an "audacious plan" (Owens, 2017).

Many others, before and since, have refused to believe it possible. Certainly, there are some serious questions to ask about its logic and philosophical assumptions (King, 2023). But those of us with long memories can recall a similar time when new technology turned the apparently impossible into routine within a mere 10 years.

Turning Point: Breaksea Island In 1976 there was a large meeting in Wellington, which discussed the problem of how to understand and manage the problem of rat damage to native fauna in New Zealand's nature reserves. Most working scientists of the time interested in rat biology and protection of wildlife were there. On the second day the group discussed available methods for managing rats, and ended with the conclusion that 'the possibility for complete extermination of New Zealand rodent populations by conventional control methods, even on islands, must be considered remote.' (Stolzenburg, 2011). This opinion was unanimous at the time, and endorsed by reputable scientists for years afterwards.

Everything changed after the invention of second-generation toxins such as brodifacoum in the early 1980s. Brodifacoum is a slow-acting toxin, which defeats the cautious taste-a-bit-to-see approach of rats to new foods, by causing no immediate ill effects. Two far-sighted scientists of DSIR Ecology Division, Rowley Taylor and Bruce Thomas, had been dismayed by the pessimistic outcome of the 1976 meeting. But they saw how the development of brodifacoum might make a significant difference, if allied with ultra-precise bait distribution ensuring that every rat had a bait station in its territory. Their success in clearing rugged, 170-ha Breaksea Island of Norway rats (Taylor & Thomas, 1989) marked a turning point in attitudes to conservation management on islands during the 1980s. It is a classic example of why conservationists, faced with apparently insurmountable odds, should never despair, because we never know what new technology might be just around the corner. But we do have to be realistic, and very patient.

There is a website that explains which pest species are monitored by DOC, and how, but it includes only ungulates, lagomorphs, and possums: https://www.stats.govt.nz/indicators/land-pests. Rats, mice, weasels, hedgehogs are important pests that cannot be monitored under the proposed scheme. Neither can the stoat, arguably the most detested of introduced predators—even though it was actually invited to come—because it is a small, energetic and very efficient killer of native bush birds and lizards, especially juvenile kiwi.

The public attitude to all invasive predators, particularly stoats, is illustrated by writer Philip Haughton. He describes how he would pace a silent bush and curse the memory of those who thought up the idea of introducing stoats to this land, because (in his opinion) deer, rabbits and opossums have not robbed us of one fraction of our heritage that the stoat has (Houghton, 1974: 122).

Houghton's concern is widely shared, and it has driven decades of efforts to get rid of stoats, none successful at national scale. But now, long-term casual observations of declining bird populations have matured into hard evidence that stoats have long had a significant impact on NZ's avifauna, and continue to pose a major challenge for control programmes (Chap. 14). One does not need to share Houghton's attitude to accept that, although stoats are indeed magnificent animals, they are efficient little killing machines that just happen to be in the wrong place. Hence the widespread public uptake of the PFNZ2050 vision.

PFNZ2050 will be a long term and hugely expensive effort. Since the time of the announcement it has blossomed into a national movement deploying more than NZ$300 million, distributed among more than 30 organisations working together in a broad collaboration of Māori iwi and hapū, NGOs, businesses, government, and community groups. They are united in a common cause and monitored by a DOC team under an official Programme Manager. It also includes a new Crown entity, Predator Free 2050 Ltd., a company set up to invest in large urban and rural projects and breakthrough research.

Some critics gave cogent reasons to challenge it in principle (Leathwick & Byrom, 2023; Parkes et al., 2017). Their immediate, considered response citing other scientists experienced in pest management was that PFNZ2050 requires a tool that kills 100% of each of the three main target species, preferably all at once. Otherwise, the effort to find and kill survivors would raise the costs of national eradication (on the mainland and on islands) to about $32 billion at then-current estimates (Parkes et al., 2017). So the massive government and private funds already committed to it need cast-iron and objective justification. Managers need a tool that can provide the best available evidence on the potentially very different outcomes of expensive alternative actions (Sutherland et al., 2004).

Progress Predator Free New Zealand 2050 is documenting inspiring results of successful pest eradication efforts on islands and at local scale on defensible parts of the mainland (Sect. 7.1 and Chap. 14). Widespread deployment by community groups armed with a certified humane trap, the DOC 200 and 250 series (Warburton et al., 2008) reassures those concerned with animal welfare as well as with efficient pest control (Littin, 2010). Certainly there are trade-offs to be considered (Reynolds, 2004), but the organised use of conventional technology is already achieving wonderful consequences for local biodiversity that are truly to be celebrated (Chap. 14).

Teams of volunteers inspired by the idea of 'restoring the dawn chorus' are doing far more for New Zealand's battered wildlife than paid staff could do on their own.

One answer to the critics might be that these objectives do not have to be applied at national scale from the start. Eradications on uninhabited offshore islands, and within fenced sanctuaries and mainland peninsulas, already meet most of these requirements, and are now routine. Applying the same experience to larger areas starts with local, defendable blocks which can be cleared, then isolated, then linked together and advanced across the country as a 'rolling front', which is how possums were dislodged from Kapiti Island (Cowan, 1992; Thomas & Taylor, 2002). But defending cleared areas on larger islands and on the whole inhabited mainland is going to prove much more difficult. PFNZ2050 is already requiring massive investment into new field and technical projects creating many jobs, against fierce competition from other applicants with different priorities, denying objections from those committed to present arrangements. Too often the real stumbling block is forgotten. Killing pests is not that hard: the main problem is preventing them from reinvading cleared areas. As Elaine Murphy once said, 'Killing stoats is like cleaning toilets: you just have to keep doing it'.

Moreover, the benefits of PFNZ 2050 are not a simple on-off switch—rather, there is a sliding scale of control impacts linked to the efficacy of protection. Trapping in small bush patches provides little benefit for kiwi, but in very large kiwi sanctuaries protecting thousands of hectares by long term trapping, adult kiwi mortality rates of 1% and chick survival rates of >60% are routine. But even there, trapping efficiency declines as the proportion of trap-shy stoats increases (Sect. 14.2.4).

The first five-year report on the PFNZ2050 programme cheerfully describes many encouraging local projects (Department of Conservation, 2021), and these are all to be heartily celebrated. But broader scale critiques are beginning to appear. PFNZ2050 in its present form lacks published evidence of systematic scientific management and detailed evidence of outcomes (Leathwick & Byrom, 2023). It is also based on the impossible assumption that mainland populations regulated from the bottom up, as stoats and rats are, can be kept suppressed *at national scale* by top-down technology alone as imposed by human action (King, 2023). Ignoring all the complex interactions among invasive species can further block recovery of indigenous fauna vulnerable to mesopredators (Norbury et al., 2013).

Justification PFNZ2050 urgently needs evidence-based proof of concept, of the kind advocated by the Centre for Evidence-based Conservation (CEBC) in UK. There is a great need for web-based support frameworks for outcome monitoring at more than local scale, and for making decisions. It is not that there is any doubt that local operations are necessary and successful; rather, it is that PFNZ2050 will eventually need to explain how it will meet its visionary ambition, *across the whole country*, despite the long-established three conditions required for any successful eradication: (1) all individuals, or at least all of one sex, must be at risk of removal; (2) they must be removed faster than they can be replaced; and (3) any future arrivals must be prevented (Parkes & Murphy, 2003).

Numbers of mustelids, like those of rabbits, are primarily bottom-up regulated by annual variation in juvenile mortality (Chap. 11 and Table 5.1). Stoat populations are naturally limited by the food available to females in winter and spring, which largely determines the number of kits born, and the number that survive to disperse (Sect. 4.7). So any effective operation that drives down the numbers of already fertilised female stoats before and during a post-seedfall surge in rodents, or removes all stoat kits before they leave the nest, can prevent a irruption of well-fed young stoats (Chap. 5). Yet, at this stage, all current official funding and research still relies almost entirely on finding ways to improve top-down methods of killing more stoats *after* they have left their nests (Murphy et al., 2019). By the time a huge new cohort of young stoats is dispersing, it is too late: the damage is already being done.

Of course, lethal methods are necessary and locally successful in the short term, to protect what is left of our surviving biota until we can think of better ways to save them, but for a long-term and larger-scale approach, we need something different. Cynics might say that, after the failed official efforts to eradicate deer and rabbits from New Zealand in the past, PFNZ2050 is a biological version of Oscar Wilde's famous description of a second marriage as the triumph of hope over experience. The passionate conservationists who turn out all round the country in their thousands and in all weathers disagree. Worse, removal of select species is not necessarily helpful. A study of predator management on 6000 ha of multi-tenure, pastoral landscape in Hawke's Bay found that removal of feral cats, hedgehogs and a few mustelids from the treatment area was offset by the opposite effect on rats (Glen et al., 2019).

A philosopher would recognise the important distinction between facts (how things are) and values (which things matter) as a means of untangling such differences. Animal welfare concerns (Littin, 2010) have certainly driven the development of more humane kill traps, like the DOC 250 and 200 series (Fig. 4.2). One off-the-wall suggestion entitled "Summoning compassion to address the challenges of conservation" (Wallach et al., 2018) was perhaps going a bit too far, except for tender-hearted people outraged by the continuing real tragedy of rats and possums trashing chicks in their nests. Surely it must be better to prevent predators and prey from meeting in the first place. The debate on how to do this continues.

13.2 Ecosanctuaries

In a New Zealand context, an ecosanctuary is defined as a project larger than 25 ha implementing multi-species, pest mammal control for ecosystem recovery objectives, and with substantial community involvement. John Innes et al. (2019) describe the attributes of 84 projects that were then meeting this definition. At that date there were three lacustrine islands, 16 marine islands, seven ring-fenced ecosanctuaries, seven peninsula-fenced ecosanctuaries and 51 unfenced mainland ecosanctuaries.

They offer last-ditch safety for a range of vulnerable species, and are successful to the extent that some island sanctuaries have actually returned threatened, previously lost endemic species to the New Zealand mainland.

In 2011 an important argument broke out in a series of Forum articles in the *New Zealand Journal of Ecology*. Paul Scofield (a Curator at Canterbury Museum), and two colleagues from Lincoln University, conducted a survey of 15 organisations responsible for fenced sanctuaries, and got replies from 12 (Scofield et al., 2011). They tabulated the costs of constructing and maintaining the fences, questioned the operators on their objectives and assessment of benefits, and surveyed alternative programmes that might gain better results for less expenditure. They concluded that fenced predator-proof sanctuaries are turning into small expensive zoos surrounded by degraded habitat that will never be able to sustain the species contained within the fence. The fact that both co-authors are from the Faculty of Commerce at Lincoln University explains the cost-benefit assumptions of this approach.

A spirited response from John Innes and six colleagues (Innes et al., 2012) acknowledges that theme in their opening words, agreeing that the focus on value for money is an increasingly important aspect of conservation management, but, as most of them are working biologists, they see the argument from a different point of view. They agree that fences are expensive, agree that they need more time to prove themselves, or at least they did at the time, and appreciate the need for careful planning. But they dispute the implication that sanctuaries try to recreate some vision of the long-lost biodiversity—rather, they are an important component of a national conservation strategy. The team reviewed the achievements of pest-free sanctuaries so far, and supported the idea of building many more of them.

I suggest that, implicit in this debate, there is an unstated but important distinction between conservation and preservation. Innes' team pointed out that sanctuaries are only part of a range of national-scale conservation initiatives, a determined *hanging on to the future,* although they also have important benefits in social engagement, public advocacy and ecotourism. The unique value of ecosanctuaries is that they are a critical component of preservation, and an essential tool for preventing otherwise likely local extinctions, that is, a *hanging on to the past.* Both are absolutely necessary.

In the same issue, Scofield and Cullen (2012) agreed with many of the same points again, but continued to demand rigorous, logical, practical methods for evaluation biodiversity projects, including fenced sanctuaries. The critical issue, then and now, is whether the conservation gains made by fenced sanctuaries over the last decade have actually justified the costs. That could not be seen then, but, more than a decade later, things are different now.

Fenced sanctuaries are becoming increasingly important to modern biodiversity protection. Yes, they are expensive, but then, so is PFNZ2050; the difference is that fences are already yielding hugely valuable results, and are much more likely than PFNZ2050 to continue doing so. Bombaci et al. (2018) show that sanctuaries can record astonishing increases in populations of endemic species (0.27–9.00 times more birds/ha than in unprotected sites), plus opportunities for translocations of red-crowned parakeet/kākāriki (*Cyanoramphus novaezelandiae*), kiwi *Apteryx* sp., robins/tātāwai (*Petroica longipes, P. australis*), hihi (*Notiomystis cincta*), kākā (*Nestor meridionalis*), whitehead/pōpokatea (*Mohoua albicilla*), NI kōkako (*Callaeas wilsoni*), and saddlebacks/tīeke (*Philesturnus carunculatus. P. rufusater*).

A very significant large meta-analysis led by Rachel Binny of 447 biodiversity responses across 16 ecological restorations quantified significant benefits for biodiversity over two decades. It confirmed that deeply endemic species had the strongest responses to relief from pest damage compared with recent native or introduced biota. It was able to predict the levels of pest suppression needed to confer significant biodiversity benefits, and concluded that the best and most economic results come from pest-free sanctuaries (Binny et al., 2020).

Obviously, the security of fenced sanctuaries depends on the integrity of the fence. Breaches of the fence are possible, so fences are closely monitored and breaches quickly detected by remote cameras (Sect. 9.5) (Fig. 13.1).

Nevertheless, accumulating experience is showing that, other than off shore islands, fences are the next best and most cost-effective available tool to protect the most vulnerable fauna from the main problem of pest control, the inevitable rapid reinvasion of cleared areas, especially by ship rats (Innes et al., 2023). It seems clear that fences should be used more often, and might be but for limitations of suitable funding and suitable sites, and objections from those who confuse welcome liberation from the risk of meeting a hungry rat with unnatural confinement.

Fig. 13.1 The largest fenced sanctuary in New Zealand (3400 ha) is protected by a fence 47 km long, surrounding Maungatautari, an eroded volcanic peak in the Waikato region. Pests constantly patrol the fence, so a breach (by falling trees or streams in spate), will probably be found and exploited within 24 h (Connolly et al., 2009). See Fig. 9.3. (Photo by Author)

There is much uncertainty inherent in lists of which native birds are most at risk of named predators, such as the one compiled by John Innes and colleagues in 2010. Despite that uncertainty (in fact, because of it) all the data we know of continue to confirm that removing all pest mammals—as in large pest-fenced sanctuaries, and building more of them—is the way forward—yet, as John says, no-one is doing this (Innes, 2024).

The problem is that small numbers of stoats can inflict a lot of damage if they ever get inside a secure location such as a fenced sanctuary. When that happens, the invader is very difficult, and very expensive, to remove. The one that invaded the Orokonui Sanctuary was eventually found in November 2015 in a den, with two kits and the remains of a kaka, by a detection dog. Orokonui had been the only area on the mainland with a resident population of South Island saddlebacks/ tīeke (*Philesturnus carunculatus*) more than 50 strong. That single incursion by a pregnant female stoat wiped them all out.

Saddlebacks/Tīeke are particularly vulnerable to stoats, as demonstrated at Tawharanui Open Sanctuary north of Auckland, which is protected by an open-ended fence across the base of a peninsula. A model technique exploring the impact of a tenfold increase in stoat detections from 2017 to 2019 on the survival rates of North Island tīeke inside the fence showed how easily even low densities of stoats can drive tīeke to extinction; by contrast, robins/toutouwai (*Petroica longipes*), and whiteheads/popokatea (*Mohoua albicilla*) were affected only marginally or not at all (Parker et al., 2024). Greater safety for sensitive species like tīeke is available only at a ring-fenced sanctuary like Brook Waimarama near Nelson, where in April 2021, a new cohort of 40 South Island tīeke was translocated from Motuara Island, in Queen Charlotte Sound.

Weasels are attracted to the smell of mice wafting through the fence from inside a sanctuary. They have an advantage in getting through tiny holes or spaces under the fence, and a huge capacity for rapid increase in such ideal conditions inside (Sect. 2.5). From 2013 to 2017 inclusive, 14 weasels were trapped inside Maungatautari Ecological Island, and one inside the Zealandia Sanctuary in 2018 (Innes et al., 2019). The Brook Waimarama Sanctuary near Nelson (700 ha) was invaded by weasels in late 2019 (at least a pair, or a pregnant female) which rapidly increased to 73 weasels trapped inside the fence by August 2020. The total numbers they reached cannot be known because then they were all removed by poison (Schadewinkel, 2021).

13.3 Reasons for Hope

Without denying the critical value of top-down methods for immediate damage control, a permanent result needs to find a way to supplement top-down logic with new tools that work from the bottom up. For success at the national level, PFNZ2050 must of course continue to develop new lethal tools, but ultimately these still feel like bailing out a bucket with a teaspoon. In the long run we need smaller buckets, not more teaspoons (King, 2023). Like a game of two halves, we must do what we can now, and also develop better strategies for the future.

To make that transition, we need to move from asking questions about how to maximise efficient killing of pests to asking different questions about how to minimise the number of pests out there to be killed. This is a wicked problem (Woodford et al., 2016), dogged by deep disagreements among stakeholders holding different world views defining acceptable technologies. Standard evidence-based approaches cannot prevent one solution from leading to other equally unsolvable ones, or may fail to resolve the issue if it is deeply influenced by layers of social inequity and history, says David Towns (2023).

Graeme Caughley used to say "If your management action does not produce the result you expected, you haven't understood the problem". That is why functional, effective management programmes for protected fauna can make progress only by first formulating clear objectives testable on real documented cases (Caughley & Gunn, 1996). By comparing the expected and actual answers to each test, researchers can be led to ask the right questions about how to protect our biodiversity at steadily increasing scales.

The Roman god Janus stood at the turn of the year, watching the transition from the old year to the new (hence the first month of the year is named January, after him). The PF2050 campaign today also stands on a significant threshold. It is suspended between past assumptions that often proved wrong, and future aspirations that cannot yet be achieved. But there is hope.

PF2050 has become a mass movement inspired and driven by community action. The people who faithfully tramp around their trap networks in all weathers share a common, unlimited vision of future restoration of threatened national identity (Glen et al., 2012). They cannot catch all the rats and mustelids that are present, and that those they do catch will be quickly replaced. But they hold on to the hope that they are helping to at least "hold the fort", and they do it for love: a much more powerful incentive than money. *Both* PFNZ2050 and fenced sanctuaries are wildly expensive: the key issue is, which strategy is most likely to protect the most vulnerable remains of our unique fauna?

Present top-down, short term strategies can provide welcome and necessary temporary relief Chaps. 11, 14), while the future bottom-up strategies that are needed for the long term, currently impossible, need some inspired lateral thinking. A contemporary version of the story of Breaksea Island could provide the turning point we need.

References

Binny, R. N., Innes, J., Fitzgerald, N., Pech, R., James, A., Price, R., Gillies, C., & Byrom, A. (2020). Long-term biodiversity trajectories for pest-managed ecological restorations: Eradication vs. suppression. *Ecological Monographs, 91*, e1439.

Bombaci, S., Pejchar, L., & Innes, J. (2018). Fenced sanctuaries deliver conservation benefits for most common and threatened native island birds in New Zealand. *Ecosphere, 9*, 02497. https://doi.org/10.01002/ecs02492.02497

Caughley, G., & Gunn, A. (1996). *Conservation biology in theory and practice*. Blackwell Science.

Connolly, T. A., et al. (2009). Estimating the potential for reinvasion by mammalian pests through pest-exclusion fencing. *Wildlife Research, 36*, 410–421.

Cowan, P. E. (1992). The eradication of introduced Australian brushtail possums, *Trichosurus vulpecula*, from Kapiti Island, a New Zealand nature reserve. *Biological Conservation, 61*, 217–226.

Department of Conservation. (2021). *Predator free 2050 5-year progress report*. Department of Conservation.

Glen, A. S., Hamilton, T., McKenzie, D., Ruscoe, W. A., & Byrom, A. E. (2012). Kiwi *Apteryx mantelli* population recovery through community-led trapping of invasive non-native mammals in Northland, New Zealand. *Conservation Evidence, 9*, 22–27.

Glen, A., et al. (2019). Predator control on farmland for biodiversity conservation: A case study from Hawke's Bay, New Zealand. *New Zealand Journal of Ecology, 43*, 3358.

Hinds, L. A., Williams, C. K., Pech, R. P., Spratt, D. M., Robinsonm, A. J., & Reubel, G. H. (2000). Feasibility of immunocontraception for managing stoats in New Zealand. *Science for Conservation, 158*, 109.

Houghton, P. (1974). *Hidden water*. Hodder and Stoughton.

Innes, J. (2024). Personal communication.

Innes, J., Lee, W. G., Burns, B., Campbell-Hunt, C., Watts, C., Phipps, H., & Stephens, T. (2012). Role of predator-proof fences in restoring New Zealand's biodiversity: A response to Scofield et al. (2011). *New Zealand Journal of Ecology, 36*, 232–238.

Innes, J., Fitzgerald, N., Binny, R. N., Byrom, A., Pech, R., Watts, C., Gillie, S. C., Maitland, M., Campbell-Hunt, C., & Burns, B. (2019). New Zealand ecosanctuaries: Types, attributes and outcomes. *Journal of the Royal Society of New Zealand, 49*, 370–393.

Innes, J., Norbury, G., Samaniego, A., Walker, S., & Wilson, D. J. (2023). Rodent management in Aotearoa New Zealand: Approaches and challenges to landscape-scale control. *Intergrative Zoology*. https://doi.org/10.1111/1749-4877.12719

King, C. M. (2019). *Invasive predators in New Zealand: Disaster on four small paws*. Palgrave Macmillan. https://doi.org/10.1007/978-3-030-32138-3

King, C. M. (2023). Asking the right questions about predator free New Zealand. *New Zealand Journal of Ecology, 47*(1).

Leathwick, J. R., & Byrom, A. E. (2023). The rise and rise of predator control: A panacea, or a distraction from conservation goals. *New Zealand Journal of Ecology, 47*(1), 3515.

Littin, K. E. (2010). Animal welfare and pest control: Meeting both conservation and animal welfare goals. *Animal Welfare, 19*(2), 171–176. WOS:000277414700006.

McDonald, R. A., & Larivière, S. (2001). The diseases and pathogens of *Mustela* spp., with special reference to the biological control of stoat (*Mustela erminea*) populations in New Zealand. *Journal of the Royal Society of NZ, 31*, 721–744.

Murphy, E. C., & Fechney, L. (2003). *What's happening with stoat research?* Fifth report on the five-year stoat research programme.

Murphy, E. C., Russell, J. C., Broome, K. G., Ryan, G. J., & Dowding, J. E. (2019). Conserving New Zealand's native fauna: A review of tools being developed for the predator free 2050 programme. *Journal of Ornithology, 160*, 883–892. https://doi.org/https://link.springer.com/article/10.1007/s10336-019-01643-0

Norbury, G. (2000). The potential for biological control of stoats (*Mustela erminea*). *New Zealand Journal of Zoology, 27*, 145–163.

Norbury, G., Byrom, A., Pech, R., Smith, J., Clarke, D., Anderson, D., & Forrester, G. (2013). Invasive mammals and habitat modification interact to generate unforeseen outcomes for indigenous fauna. *Ecological Applications, 23*, 1707–1721.

Owens, B. (2017). The big cull: Can New Zealand pull off an audacious plan to get rid of invasive predators by 2050? *Nature, 541*(7636), 148–150. WOS:000396125500013. http://www.nature.com:80/polopoly_fs/1.21272!/menu/main/topColumns/topLeftColumn/pdf/458148a.pdf

Parker, K. A., Lovegrove, T. G., Maitland, M., Parlato, E., Stone, Z., & Armstrong, D. P. (2024). Quantifying the impacts of stoat and cat incursions on reintroduced bird populations in a predator-fenced wildlife sanctuary. *Animal Conservation*. https://doi.org/10.1111/acv.13000

Parkes, J., & Murphy, E. (2003). Management of introduced mammals in New Zealand. *New Zealand Journal of Zoology, 30*, 335–359.

Parkes, J. P., Nugent, G., Forsyth, D. M., Byrom, A. E., Pech, R. P., Warburton, B., & Choquenot, D. (2017). Past, present and two potential futures for managing New Zealand's mammalian pests. *New Zealand Journal of Ecology, 41*, 151–161.

Reynolds, J. C. (2004). Trade-offs between welfare, conservation, utility and economics in wildlife management—A review of conflicts, compromises and regulation. *Animal Welfare, 13*, S133–S138.

Russell, J. C., Innes, J. G., Brown, P. H., & Byrom, A. E. (2015). Predator-free New Zealand: Conservation country. *Bioscience, 65*, 520–525.

Schadewinkel, R. (2021). Personal comunication.

Scofield, R. P., & Cullen, R. (2012). Fenced sanctuaries need critical evaluation: A reply to Innes et al. (2012). *New Zealand Journal of Ecology, 36*, 239–242.

Scofield, R. P., Cullen, R., & Wang, M. (2011). Are predator-proof fences the answer to New Zealand's terrestrial faunal biodiversity crisis? *New Zealand Journal of Ecology, 35*, 312–317.

Stolzenburg, W. (2011). *Rat Island: Predators in paradise and the world's greatest wildlife rescue*. Bloomsbury.

Sutherland, W. J., Pullin, A. S., Dolman, P. M., & Knight, T. M. (2004). The need for evidence-based conservation. *Trends in Ecology & Evolution, 19*, 305–308.

Taylor, R. H., & Thomas, B. W. (1989). Eradication of Norway rats (*Rattus norvegicus*) from Hawea Island, Fiordland, using brodifacoum. *New Zealand Journal of Ecology, 12*, 23–32.

Thomas, B. W., & Taylor, R. H. (2002). A history of ground-based rodent eradication techniques developed in New Zealand, 1959-1993. In C. R. Veitch & M. N. Clout (Eds.), *Turning the tide: The eradication of invasive species* (pp. 301–310). IUCN SSC Invasive Species Specialist Group.

Towns, D. (2023). *Ahuahu. A conservation journey in Aotearoa New Zealand*. Canterbury University Press.

Wallach, A. D., Bekoff, M., Batavia, C., Nelson, M. P., & Ramp, D. (2018). Summoning compassion to address the challenges of conservation. *Conservation Biology*. https://doi.org/10.1111/cobi.13126

Warburton, B., Poutu, N., Peters, D., & Waddington, P. (2008). Traps for killing stoats (*Mustela erminea*): Improving welfare performance. *Animal Welfare, 17*, 111–116.

Woodford, D. J., Richardson, D. M., MacIsaac, H. J., Mandrak, N. E., van Wilgen, B. W., Wilson, J. R. U., & Weyl, O. L. F. (2016). Confronting the wicked problem of managing biological invasions. *NeoBiota, 31*, 63–86. https://doi.org/10.3897/neobiota.3831.10038

Part V

Part V Judgement

14 Assembling the Evidence

Abstract

Evidence against predation as a driving cause of population decline must be able to distinguish between different and interacting causes of loss. Predation can be blamed only if it removes individuals that might otherwise have lived. Detailed accounting of all causes of mortality of a given threatened species has been possible only since the development of the advanced technology summarised in Part IV. Control of stoats can benefit all native species generally, but their effects are usually confused with those of rats. The best evidence comes from rat-free Resolution, Secretary, and Anchor Islands, where stoats eradicated resident populations of Kākāpō (*Strigops habroptilus*), SI Kōkako (*Callaeas cinerea*), SI piopio (*Turnagra capensis*), S I tīeke (*Phiesturnus carunculatus*), kākāriki (*Cyanoramphus* sp.), and little spotted kiwi (*Apteryx owenii*). Stoats that invaded Maud Island and Orokonui fenced sanctuary wiped out formerly thriving refugee populations of saddlebacks/tīeke. Suppression of stoats on Waiheke Island has greatly increased nesting success and numbers of kākā (*Nestor notabilis*). Ferrets are deadly to larger birds, including breeding adult kiwi, weka (*Gallirallus australis*) and accessible seabird colonies, but are poor climbers and have never reached the most valuable islands. Weasels are generally small and now rare, but were very destructive to small prey on first arrival.

Keywords

Island refuges · Fenced sanctuaries · Impacts of invasive mustelids · Mortality accounting · Predation risk to endemic species · Refuge islands · Early abundance of weasels · Tutira station

C. M. King, *Stoat in the Dock*, Fascinating Life Sciences,
https://doi.org/10.1007/978-981-96-4562-6_14

Predation and habitat fragmentation are generally agreed to be the two main causes of damage to the native biota of New Zealand. Habitat loss is bad, although mitigated to some extent by a respectable national coverage of well-established protected natural areas, but they are not enough. Understanding predation is a more dynamic problem, because it depends on rigorous analyses and numerical models of many different shifting ecosystems. These need field-based, quantitative measurements capable of accounting for the extent of mortality definitely due to each species of predator relative to all forms of natural mortality or failures in recruitment. Over the last few years, intensive field studies allied with computer models have begun to distinguish between their effects.

Such analyses have been possible only since the development in the late 1980s of techniques able to measure productivity and mortality of eggs and chicks, and identify their causes (Chap. 11). The general conclusion is that habitat protection is the first and absolutely necessary step, but, as Takahē Valley proved in the 1970s (Sect. 3.2) it is not by itself sufficient. There, as in many other places, invasive predators are only some among other prime suspects. Is this true?

A rational judgement needs to examine the case against stoats, weasels and ferrets with a suitably rigorous review. In a different context, the legal system has developed over centuries as the best and most objective way to decide between conflicting explanations of serious crime. In a court case, both sides have to know their subject intimately, because every statement they make to the court can be instantly challenged in minute detail by the other side. The rules of presentation of evidence in court are always applied objectively and immediately. The judge and all the barristers must be qualified and experienced in the legal issues concerned, and in the required court procedures. The tightly restricted and time-limited format of a legal action helps to define the boundaries of the debate, and to pour cold logic on all interpretations other than those that survive the immediate challenge of the other side. It sounds like a useful model to apply here. My title, *Stoat in the Dock*, tries to reflect that process.

14.1 Cautions

History warns us that attempts to rescue a threatened species by focussing on a single cause of losses such as predation does not necessarily help it, because the dynamics of both predator and prey are ruled by a complex web of interacting community processes. Moreover, predator control is only a part of the wider aim of biodiversity restoration, whose success also depends on resolving wider social, economic and cultural issues (Norton et al., 2016; Towns, 2023).

But where predation can be objectively identified as a process of real concern, the next stage is to calculate its impact on a given target population. This is largely a matter of juggling relative numbers and rates of change, both among the predators concerned and their prey. The evidence available ranges in quality from (1) the merely suggestive, through (2) circumstantial to (3) conclusive.

Suggestive only studies include the great majority of predator diet reports, derived from simple gut or scat analyses with no supporting information about the prey. They offer valuable first steps helping to focus concentrated attention on the species or locations of concern, but do not themselves offer any information on predation rates.

Circumstantial or contributary evidence may add important details such as the proportion of a protected species killed by stoats, and what other factors could contribute to their losses, but cannot always distinguish between them or exclude other explanations. Habitat modification, food deprivation or loss of safe nesting areas are at least as important as predation in explaining the damage to native biodiversity, and in places more so, but where these are studied separately, their important interactive effects cannot be understood (Didham et al., 2007; Towns, 2023).

Populations of native birds can also be drastically affected by variations in their main food supply or its alternatives, because the risk of a bird failing to find sufficient supplies of suitable food every day can often be higher than the risk of meeting a predator. The distribution of native species is also determined by external variables like geography and climate; competition with other native species; and predators other than mustelids. These first two categories cannot provide the hard evidence needed, but can help lead to it.

Conclusive evidence secure enough to lead to a firm and practical management policy can come only from proof of a real (or realistically modelled) sustained increase in the numbers and distribution of the protected species. Even then it has to be due *only* to documented improvements in the productivity and density of breeding adults and survival to breeding age of chicks, and *only* where mustelids are removed faster than they can replace themselves *and* are prevented from reinvading. Outcome monitoring of such quality and at landscape scale is time-consuming and expensive. Understandably, such evidence is rare, and not guaranteed to follow even successful predator removal (Coté & Sutherland, 1997), but where available it is included in the following species accounts.

Since the three mustelids in New Zealand differ in their predatory activities, the cases against them have to be considered separately.

The arguments advanced here are based on three fundamental assumptions:

1. The fates of populations cannot be deduced from observations of individuals.
2. Any agency attempting to control a target population artificially must first understand what controls it naturally.
3. Biodiversity outcome monitoring must be able to document real benefits to the protected species in a specified landscape, not merely as a list of dead predators.

The general public, and many conservationists, need to appreciate these important but counter-intuitive ideas, so it is worth spelling them out here.

14.2 The Case against Stoats

Stoats are as agile in the canopy as any squirrel, so few arboreal prey are out of their reach. They are small enough to enter the burrows of rabbits and rats, or even the dome nests of small birds, and sinuous enough to turn round inside. Their sharp claws and light weight enable them to climb up tree-trunks and lichen-covered rocks with ease, jump across considerable gaps, and run down again head first. They can climb narrow saplings, and cross between adjacent trees by running along connecting branches; they are sometimes caught in arboreal rat traps (Sect. 3.1). Hence, forests certainly hold more prey than does pasture for a stoat.

Prey are detected by sound, sight (especially if moving), or smell. Broods of chicks, especially if calling, are taken more often than clutches of eggs (Sect. 14.2.3) because stoat hearing is acute and directional. Stoats swim readily and well, in both fresh and salt water, and could potentially reach offshore islands within up to c. 5 km of the mainland coast if assisted by tides, currents or floating debris (Sect. 7.2).

The most convincing evidence comes from cast-iron records of what stoats can do in the absence of other mammals. On rat-free Resolution, Secretary, and Anchor Islands, stoats wiped out resident populations of Kākāpō (*Strigops habroptilus*), South Island (SI) kōkako (*Callaeas cinerea*), SI piopio (*Turnagra capensis*), SI saddleback/tīeke (*Phiesturnus carunculatus*), kākāriki (*Cyanoramphus* sp.), and little spotted kiwi (*Apteryx owenii*). Similarly, saddlebacks/tīeke were wiped out when stoats got to Maud Island, and in Orokonui fenced sanctuary (Sect. 13.2), and were drastically reduced in Tawharanui and Shakespear Regional Parks. Shore plovers/tuturuatu (*Thiornis noveseelandiae*) were suddenly at serious risk when stoats got to Motutapu (Veale, 2024).

On the other hand, stoats are less abundant than hedgehogs, ferrets or cats per unit area in pasture (Keedwell & Brown, 2001), which is by far the dominant land use of the two main islands. Stoats avoid being exposed to the scrutiny of harrier hawk/kahu (*Circus approximans*) and falcons/kāeaea (*Falco novaeseelandiae*) searching for prey from the air, and pasture holds fewer prey in a roughly one-dimensional habitat compared with the 3D environment of forests. Even so, systematic monitoring by remote cameras (Breedt & King, 2021) shows that a few stoats per hectare of New Zealand's vast pastoral landscape adds up to a larger total population of stoats there than in smaller areas of forest where stoat density can be higher (Fig. 9.2). A sparse but very widely dispersed population of stoats is a permanent threat to protected forest patches easily invaded from surrounding pastoral land (Sect. 11.2). During mouse peak years stoats often disperse above the tree line into tree-less alpine habitats, or across water to inshore islands. These characteristics of stoats explain why so many native species are at risk.

14.2.1 South Island takahē

SI takahē (*Porphyrio hochstetteri*) are large, flightless herbivorous birds once widespread across much of lowland southeast South Island, but now largely dependent on the thick alpine tussock grass clothing their last high alpine refuge. Their North Island equivalent species is already extinct. In summer, takahē forage by plucking out tussock grass stems individually, holding the stem in one foot, and using their powerful beaks to strip out the nutritious material from base of the stem (Fig. 14.1). As snow covers the tussock in winter, they retreat to nearby beech forest, where they are vulnerable to stoat irruptions after a seedfall.

In 1948, the recently rediscovered SI takahē were granted rigorous protection in a 518-km^2 'Special Takahē Area', closed to all but official human access.

Fig. 14.1 A South Island takahē holds a tussock grass in one foot while extracting nutritious juices from the base. (Photo by Author)

Conservation theory at the time considered that protection from human interference would be sufficient to save them, but they continued to decline (Tennyson, 2010). By the late 1960s their failure to recover was being attributed to stoat predation. The limited trapping operation started then, and the introduction of a new trapping regime in 1972 (Sect. 3.2), could not prevent a further decline to a low of 120 birds in 1981. Safeguard groups have been transferred to offshore islands, where they play an important part in the species recovery plan even though their breeding success is often poor (Lee & Jamieson, 2001: 111–113).

Adult takahē have massive beaks, and one at least is known to have actually killed a stoat (Lavers & Mills, 1978). Most takahē do not fight back, so a further intensified stoat trapping campaign in 1992–2008 improved the annual survival of adult takahē from 78% to 85%, and of chicks from 75% to 83%. Losses of fertile eggs (usually two per clutch) are apparently low (3–5%), but young chicks are vulnerable. Unprotected nests fledge only an average of 0.43 chicks a year. A massive stoat irruption during the winter of 2007 almost halved the Fiordland population (from 168 to 93 known individuals) (Hegg et al., 2012).

After millennia of adapting to a safe, pre-human life in New Zealand, takahē are no longer good at defending their chicks from danger. By contrast, the pūkeko (*Porphyrio melanotus*), a more recent immigrant from the same Australian stock as the ancestors of the takahē, teaches its chicks to understand and respond to the threat of an approaching stoat. One cross-fostered takahē chick learned this behaviour from its pūkeko guardians, while a parent-reared takahē chick did not (Bunin & Jamieson, 1996). A captive rearing programme feeds young chicks with takahē-head puppets (to avoid imprinting the chicks on their human carers), and subadults are tutored in foraging techniques by older buddies (Lee & Jamieson, 2001: 93).

The estimated probability of a viable egg producing a 1-year-old bird is 65.0% in captivity, but only 27.6% in the wild.

On the other hand, stoats are not the only threat to the survival of these ancient birds. The takahē was once found throughout the South Island, but is now confined to a limited refuge in the Murchison Mountains of Fiordland, also invaded by red deer. Takahē have managed to survive in a severe alpine climate, where cold winters and heavy snowfall significantly increase their overwinter mortality rate. Both takahē and red deer are able to distinguish the most nutritious tussocks and preferentially forage on them, so competition between birds and deer is minimised by regular culling of deer.

The long term survival of takahē depends on a combination of management strategies, which certainly has to include both captive rearing at the advanced breeding centre at Burwood Bush near Te Anau, and established reserve takahē populations on islands and in 17 scattered ecosanctuaries, but neither would be enough without extensive, ongoing stoat trapping in the Murchison Mountains. Recent or planned new release sites in the Kahurangi Range in northwest Nelson, and in the Greenstone and Rees Valleys at the head of Lake Whakatipu, will put the birds in the reach of ferrets and feral cats for the first time. Their future will demand careful watching of how takahē respond these new challenges (Innes, 2024).

14.2.2 Mohua and Other Beech-Forest Passerines

Mohua/yellowheads (*Mohoua ochrocephala*) are small hole-nesting birds that can best live and breed in the high canopies of mature red beech trees, but their nests are not out of reach of post-seedfall irruptions of rats (Fig. 14.2). They were once very abundant and widespread, but have now disappeared from more than 75% of their former range. Mohua produce a second brood later in the season than do many birds, whose sitting females and fledglings are most vulnerable just when the annual crop of young stoats is dispersing in December–February (Elliott, 1996b).

Stoats evolved in the northern hemisphere, where their favourite prey, voles, field mice and lemmings, are meatier and more accessible than are birds, especially in winter. Even in New Zealand, stoats often prefer to hunt rodents rather than birds when possible (Sect. 5.4). In their native habitats, when rodents are abundant after a masting season producing huge crops of acorns and beechnuts, northern birds are buffered against mustelid predation (Sect. 2.2). Alas, in New Zealand, masting events have the opposite effect.

In the feast or famine conditions of the high Arctic, mustelid biology is closely linked to population cycles of small rodents, so they are pre-adapted to flourish during mouse peak summers. But the consequences of mast seeding depend not only on those species most directly affected by the masting event, but on the species composition and interrelationships of the total ecosystem (White & King, 2006). The very much smaller species assemblage of small mammals and birds in New Zealand beech forests is part of the reason that mast-seeding events have such drastic consequences for birds here.

Fig. 14.2 Prey remains retrieved from a rat nest included feathers and leg bands from a colour-banded mohua, plus rat pellets. (Photo by Colin O'Donnell)

The stoat's two key prey items in its native habitats (McDonald et al., 2000) are absent (voles) or scarce (rabbits) in beech forests. Wild house mice (*Mus musculus*) are small (most local mean weights vary from 18 to 25 g), and much less meaty than voles, so are not an adequate substitute. They cannot normally protect native birds against extra predation in post-seedfall years, as is common in the northern hemisphere. There is only one recorded occasion when that happened (Sect. 5.4). In a typical average mast year, followed by increases in mouse abundance indices of up to around 25%, stoats increase in numbers up to five fold, but their guts contain about the same proportion of birds as usual (King, 1983). The same applies to ship rats, so more stoats and rats all eating the usual numbers of birds per head means more birds lost. Then, mouse peaks crash while stoats are still abundant, precipitating high mortality among newly independent young stoats. If by then few birds are available, some adult stoats can survive on invertebrates (Sect. 4.3).

Stoats and rats are confident and very agile climbers, so they can both easily reach the mohua nests even at great heights above the ground. During mouse peak years, the productivity and survival of female mohua is greatly reduced (only the females incubate), and fledgling success is seriously damaged because larger than usual numbers of stoats and rats can kill all the chicks plus the sitting females, which have no escape from the nest hole (Elliott, 1996b). Both stoats and rats carry their kills back to their dens (Fig. 14.3), where they can separate the meat from the feathers and legs (Fig. 14.2).

So the additional numbers of mice available following a mast year permit a five- to sixfold increase in stoat populations without diverting the stoats' attention from

Fig. 14.3 Rock wrens' nests are seldom out of reach of agile stoats. Here a stoat jumps across a 1 m wide gap across a rock face carrying a chick. (DOC remote camera footage, analysed by Little et al., 2017)

eating birds, including threatened species such as mohua. The damage is due to increased numbers of stoats mostly taking the usual number of birds each, rather than to the more conventional explanation of prey switching to birds by a stable number of stoats. In contrast to predictions from northern-hemisphere studies, stoat irruptions lead to additional predation in the *same* summer that the mice were abundant, not in the autumn after the mice had gone (Sect. 5.4).

By 1996, enough had been learned about the effects of stoats on mohua to support an integrated Hole-Nesting Bird and Predator Programme. A special issue of the *New Zealand Journal of Zoology* for Volume 23 in 1996 was dedicated to a series of papers describing this work (O'Donnell, 1996). For example, stoat numbers, mohua productivity and adult female mortality were compared in two study areas, one trapped and one untrapped, in the Eglinton Valley, through a stoat irruption in summer 1990/91 and in the summers of 1991/92 and 1992/93 when stoat numbers were lower. The fledging of many first clutches, and the laying of second clutches, normally coincides closely with the period when high numbers of stoats are being caught in traps. Eighty percent of the nests in the trapped area fledged young, compared with only 36% in the untrapped area. Pairs produced nearly twice as many young in the trapped area. A higher proportion of breeding females disappeared from the untrapped area.

These results suggest a valuable benefit of predator control for mohua populations (Dilks, 1999; Dilks et al., 2003). The long-term consequences for the prospects of extinction of a given mohua population depend on the relative frequency of good years that permit mohua to raise two broods in a season (Elliott, 1996a), but it takes very few stoats to cause a lot of damage to a breeding population (Choquenot, 2006).

On the other hand, part of the reason that trapping through the 1990/91 stoat irruption could successfully protect mohua in Fiordland was that ship rats were then scarce in the Eglinton Valley. Rats have invaded that area since 1995, so now, the same traps that kill stoats also kill rats, but not nearly enough of them. During the massive post-seedfall event of 1999–2000, rats killed more than 80% of the local study population of mohua (O'Donnell, 2024). Un-replicated trap results and models based on data from mainland areas where stoats and ship rats co-exist cannot distinguish between their effects without parallel data from untreated areas.

14.2.3 Rock Wren

Rock wrens/piwauwau (*Xenicus gilviventris*) are small, semi-flightless passerines confined to the high alpine regions of the South Island. They are especially precious because they represent an ancient but much diminished family of birds (5 of 7 species already extinct) that has been distinct from all other passerines since the Miocene (Worthy et al., 2010). Once common throughout New Zealand, rock wrens have been declining for many decades (Michelsen-Heath & Gaze, 2007).

Stoats are well adapted to hunting in cool rocky climates, because they originated in similar habitats in the boreal tundra of the northern hemisphere, whereas ship rats, originally a tropical species, are normally scarce above the tree line. New Zealand alpine grasslands are still a good-quality habitat for stoats even in mouse-poor years because they can, when necessary, survive on invertebrates alone (Smith et al., 2008).

In 2012–2015, DOC's remote cameras observed stoats or mice (but no rats) visiting 13 of 32 often inaccessible nests of the endangered rock wren (Fig. 14.3). Nests were most frequently attacked during the day (85% of nests) and at the chick stage, because stoats have directional hearing and could locate the source of chicks calling for food. Stoats sometimes looked straight at the camera lights and could clearly see them (Sect. 10.2), but ignored both them and the nest visits by field workers. Adult stoats mostly hunted alone, taking chicks, whole clutches of eggs and adults (Little et al., 2017).

In 2015, a network of 89 stoat traps was established in the Haast Range and 37 in the Homer-Gertrude Cirque. Daily survival rates, egg-hatching and fledgling rates were all improved by trapping. Overall, the survival rates of rockwren nests ranged from 2% to 28% at unprotected sites, and from 47% to 67% at sites protected by trapping (Weston et al., 2018). The rock wren population in four study areas increased from 100 to 250 adult birds.

An organised programme in 2016 confirmed the real benefits of aerial 1080 operations (over 694,000 ha) for rock wrens during the post-seedfall years when they are at special risk from stoat and rat irruptions (Chaps. 5 and 8). Rock wren nesting success in the monitored areas of the 1080 drop zone more than doubled, to >50% compared with <25% in untreated areas (Elliott & Kemp, 2016).

Assuming that the absence of rats from the nests observed was true for rock wren habitat in general, at least in those years, these nesting data supported by a sustained increase in wren numbers are convincing for now. Lack of food and cool temperatures keep ship rats away from alpine areas, except after a mast year when massive extra food allows a temporary surplus of rats to invade the higher elevations (Carpenter et al., 2022).

Past climate changes have also affected the habitat of other alpine species such as the kea, and presumably also have had consequences for the rock wren. Future climate change will disadvantage all alpine specialists if the tree line is raised enough to limit the extent of their only environment. Warming winters also allow ship rats to spread upwards into alpine areas that were previously too cold for them, increasing the effects of 'thermal squeeze' on sensitive alpine species (Harris et al., 2022).

14.2.4 Kiwi Chicks

New Zealand has five living species of kiwi, all endemic at Family (the highest possible) level, hence with supreme value in both scientific and national senses. They were all formerly abundant on the mainland, but are now reduced to remnants. The North Island brown kiwi *Apteryx mantelli* (Fig. 14.4) retains five relatively good but widely separated populations; the great spotted kiwi (*A. haastii*) is confined to northwest Nelson and nearby ranges; the southern tokoeka (*A. australis*) is still present in parts of Westland, Fiordland and Stewart Island; and the Okarito brown kiwi (*A. rowi*) clings on to a tiny refuge in central Westland.

Fig. 14.4 A radio-tagged brown kiwi chick, safe inside the fence around Maungatautari Sanctuary Mountain. (Photo by Author)

These four kiwi species are all vulnerable to stoat predation. Only the fifth species, the little spotted kiwi (*A. owenii*), is never likely to meet a stoat because it is confined to five predator-free offshore islands. Norway rats once swarmed across the mainland and many islands, when they destroyed many kiwi long before stoats arrived (Douglas, 1894: 71–74), but they have since been replaced by ship rats, which do not (King et al., 2011).

DOC's Fourth Kiwi Recovery Plan 2018–2028 (Department of Conservation, 2018) sets a national target of at least 2% annual population increase to 100,000 kiwi by 2030. Figure 14.6 of that plan shows that five of nine named kiwi populations are achieving that growth rate, and in all five, 50% or more of the birds live under existing management. Four others with 20% protection or none are still declining. So the gains made by the protected populations are being offset by the losses, especially due to stoats in the South Island. The present overall growth rate of 1% a year is much too slow to reach the 2% threshold.

Kiwi eggs are very large and laid in a burrow, incubated and defended by the male. Kiwi chicks can feed themselves soon after hatching, but remain close to their natal nest site at first. Chicks under 800 g or so in body weight are very vulnerable to predation, especially by stoats; after that, they can defend themselves nearly as well as adults (McLennan et al., 2004). In Trounson Kauri Park during a comprehensive predator control programme in 1996–2002, the survival of 92 monitored kiwi chicks was increased from a normal average of <10% to 38% overall. Many of what kills were recorded could have been the work of a few individual stoats (Gillies et al., 2003).

A landmark review by John McLennan's team of the extent of predation on damaging unprotected kiwi populations throughout the country found that, up to 1996, only 14 out of 209 adult brown kiwi died during 159.6 radio-tracking years, mainly due to ferrets or dogs. Sixty-nine (68%) of 102 eggs from 77 nesting attempts by 48 pairs failed to hatch, most destroyed by brushtail possums (*Trichosurus vulpecula*) or mustelids. The most shocking part of their report was their confirmation that very few—probably only three of 49 chicks—survived to adulthood, a juvenile mortality rate of about 94%, most attributed to stoats or cats. Overall, predators killed at least 8% of chicks, 45% of juveniles, and possibly as many as 60% of all young kiwi, leading to an estimated decline of kiwi numbers in mainland forests of almost 6% a year (McLennan et al., 1996). In Northland, unmanaged kiwi populations are still declining at 2.5% a year (that is, their numbers halve every 27 years) (Robertson et al., 2011).

Despite these losses, kiwi have survived, largely because the adults are normally long-lived and can breed again over many successive seasons. But to retain a local kiwi population, McLennan's model predicted that chick survival and recruitment must be raised to at least 19%, which requires that stoat numbers must be reduced over large areas to <2/km^2 (i.e., by 80% in stoat plague years), and maintained that low for up to 9 months (Basse et al., 1999). This is a tall order because, in the summer of a good mouse year, stoat densities can reach up to an estimated 10 per km^2. Another study suggests that a supposed safety threshold of <2/km^2 is, if anything, too high (McDonald & Murphy, 2000).

Between 2003 and 2009, the survival of Fiordland tokoeka chicks was higher in trapped areas (37%) than in non-trapped areas (17%). Regular trapping can produce a stable or very slightly increasing kiwi population (by 1.1% a year) (Edmonds, 2015). But in the absence of stoat trapping, the tokoeka population is predicted to continue to decline slowly (by −1.3% a year).

An effective local reduction in the numbers of stoats can produce increases in counts of kiwi calls, a proxy measure of kiwi abundance (Pierce & Westbrooke, 2003), and real improvements in the productivity and survival rates of juveniles. In the trapped parts of one study area beside Lake Waikaremoana, recruitment of kiwi chicks into the adult population reached 17 of 30 monitored chicks (58%). In comparison, in non-treatment areas only 2 of 26 monitored chicks (7%) were recruited by the third year of the study (Blackwell et al., 2003).

The fatal effects of stoat predation on kiwi chicks, and the difficulty of removing enough stoats to prevent it, are illustrated by a well-documented case in which even intensive stoat trapping failed to protect 14 known and monitored kiwi chicks. In a protected forest at Okarito, predator trapping killed >10,000 rats and 1950 stoats between April 2001 and July 2004. Chick monitoring in the summer of 2001–2002 found 30% chick survival, but after the heavy podocarp masting season of 2002, 13 radio-tagged chicks were killed by stoats, and one died for unknown reasons (Murphy et al., 2008).

The problem is that annual stoat trapping cannot remove untrappable individuals, that is, the cautious ones (Johnstone et al., 2023). These can continue to kill chicks, unless an occasional use of 1080 in kiwi-proof tunnels removes trap-shy stoats and improves survival of the remaining chicks (Robertson et al., 2016). Rats do not attack kiwi chicks, but they support stoats that do. In very large kiwi sanctuaries protecting thousands of hectares by long term trapping of both stoats and rats, adult mortality rates of 1%, chick survival rates of >60% and population growth rates of 6% are routine. But even there, trapping efficiency declines as the proportion of trap-shy stoats increases. The progressive reduction in trapping effectiveness is well illustrated at Whangarei Kiwi Sanctuary, from 67% in 2003 to 24% in 2008, and at Moehau, from 95% in 2002 to 38% in 2007 (Robertson & de Monchy, 2012). Large and important offshore islands cannot be cleared of stoats by trapping alone.

The conclusion that "the persistence of kiwi on mainland New Zealand is now largely dependent on the development of new techniques for controlling stoats" (Basse et al., 1999) was one of the driving forces behind PFNZ2050, and in much of the South Island it is still true. At the moment, suppression of stoats by regular trapping on a large area is possible, but total eradication is not, even on uninhabited Fiordland islands (Chap. 7), because some adult stoats simply refuse to engage with any human device (Veale et al., 2013).

Part of the puzzle is that kiwi populations are also limited by other factors independent of stoat numbers. The minimum viable area for a self-sustaining kiwi population must be large, at least 100 km^2, but protecting chicks by suppressing stoats over such large areas is difficult (Basse & McLennan, 2003). Forest clearing and fragmentation have reduced kiwi living spaces to patches mostly smaller than that, so surviving chicks often have to emigrate to or through unsafe areas. Moreover, stoat control cannot protect chicks lacking profitable foraging areas.

So the best prospects for kiwi are in places protected by natural barriers to emigration, such as in the Moehau kiwi sanctuary at the tip of the Coromandel peninsula. Elsewhere, the ultimate goal of controlling stoats to help the surviving and often aging breeding adult kiwi replace themselves has been hard to achieve. One study on Whangarei Heads, where in 2011, call rates of kiwi living there were the highest recorded since 2001, concluded that any such real increases in numbers of kiwi remains hard wherever stoat abundance exceeds a catch per unit effort threshold of ~0.1 stoat *per trap per year* (Glen et al., 2012).

Other causes of kiwi mortality include the loss of breeding stock (Sect. 14.3.1), especially to dogs and ferrets, and, less frequently, road kills, leg-hold traps, the burden of carrying a transmitter, cyanide set for possums, and invertebrates affected by 1080. Fortunately, all that bad news is not the end of the story.

The fightback has been led by multiple community organisations, which are now making real progress. In April 2024, Savethekiwi.nz released the results of an inhouse research project that outlines the estimated growth of the North Island brown kiwi population. They listed the numbers reported by 25 of the projects they fund, in 2019 and again in 2023, and analysed the difference (For updates, check info@savethekiwi.nz) The results show that, in combination, these projects have increased the population of North Island brown kiwi by approximately 7000 birds. These results were possible only because of the funding and the work that Māori- and community-led groups have been able to put into it. While this is heartening news, there is still a lot of work to be done to take kiwi from endangered to everywhere, but progress is impressive.

To take just one example, kiwi had disappeared from the Wellington area some 200 years ago. But the city stands on a broad peninsula facing south, so local conservation group Capital Kiwi could operate intensive trapping programmes (4600 traps over 23,455 ha) with minimal risk of reinvasion of pests except from the north. Hundreds of kiwi have been transported into areas were, 25 years ago there were none (Peck, 2024).

14.2.5 North Island kōkako

The NI kōkako (*Callaeas wilsoni*) is a strongly endemic forest bird that once occupied nearly all the extensive native forest of the North Island (Fig. 14.5). As forest fragmentation has proceeded, NI kōkako have gradually disappeared from the remaining smaller forest patches (under 1000 ha). By the early 1980s, surviving populations were in poor condition. For example, at Mapara (1400 ha) in 1980/81, there were 16 pairs of adults, which produced one brood per season; c. 95% nests were lost to predators; the mean productivity was 0.13 chicks/pair/season, total $16 \times 0.13 = 2$ chicks/year; the mortality rate of adults was c. 10%, requiring replacements of 10% of 16 = 1.6; the overwinter survival of fledglings was unknown but

Fig. 14.5 The North Island kōkākō. (Photo by Bernie Krippner)

not high. The Mapara birds were on the margin of failing to cover their losses, and were continuing to decline slowly (Innes & Hay, 1995).

The predators contributing to these losses were not identified, but stoats were assumed to be among them. Kōkako were also vulnerable to competition for food with brushtail possums, an introduced browser that arrived in the central North Island after the 1970s, and soon began to damage the food supply of fruit-eating native birds. For example, all 13 kukupa (NZ pigeon) nests in Mōtatau Forest (350 ha) started in late 1997 failed at the egg stage, but after intensive trapping of possums and ship rats, losses were reduced to less than 4%. Seven nests located in 1998–99 successfully fledged young. Counts of native pigeon/ kererū/kukupa (*Hemiphaga novaeseelandiae*) adults subsequently increased (Innes et al., 2004).

The NI kōkako lives mostly in non-beech forests, and has survived into the age of intensive conservation management. Why then could the South Island kōkako (*C. cinerea*) not do the same? One suggestion is that the South Island subspecies lived in beech forests, so it could have been an early casualty of the periodic increases in predation by stoats after beechmast years (Clout & Hay, 1981).

To identify whether browsers or predators should be most urgently removed to protect kōkako, a visionary 8-year experiment was designed to disentangle their effects. Browsers or predators were switched on and off separately, and kōkako productivity was compared with that on pest-free Little Barrier Island. After the first year or so, stoat trapping was discontinued. The conclusion was that, even without stoat trapping, kōkako productivity recovered when *and only when* possums and rats are removed to specified threshold levels (Innes et al., 1999). This does not mean that stoats were irrelevant, only that they could more efficiently be removed by secondary poisoning aimed at possums and rats (Chap. 8).

Productivity per female increases in response to the reduced number of rats and possums. The estimated fledging rate necessary to stabilise the kōkako population increases predictably up to some threshold, at which behavioural processes (territoriality, competition) become more important (Sinclair et al., 2006).

Effective removal of ship rats and possums is a key first step that has enabled NI kōkako to reach 2000 pairs by 2021. Sanctuary Mountain Maungatautari (3400 ha) is the largest of the North Island sanctuaries, free of all invasive mammals except mice. The growth of the kōkako population translocated there is the fastest ever seen.

Elsewhere, stoats are suspected to be important intermittent predators, especially of subadults and adults, and they are more dangerous than rats and possums because they can kill these birds year-round while they are roosting, rather than just when they are nesting. One of the largest translocations attempted (to Secretary Island) where eradication of stoats has so far failed (Sect. 7.3), lost all adult kōkako in a couple of years. Most kōkako recovery sites have targeted stoats in most years, but much remains to be learned about stoat impacts and control (Innes, 2024).

14.2.6 Whio

Whio (*Hymenolaimus malacorhynchos*) are native torrent ducks confined to fast flowing rivers in mountain country (Fig. 14.6). Their linear habitat offers a highway to their nest sites along river banks, which are frequently patrolled by stoats and Norway rats. In beech forest river systems, whio are especially vulnerable to post-seedfall irruptions of stoats. Amy Whitehead's study in Fiordland compared the number of fledglings per pair of adults across a range of mean annual stoat abundance indices from 0% to 90%. In five areas where the stoat abundance index was 0–2%, fledging success ranged from <0.4 to 1 chick per pair. In four other areas with stoat abundance indices ranging from <5% to 90%, fledgling success was zero. The margin of difference was very slim, showing that stoats are very efficient at finding whio nests and fledglings (Whitehead et al., 2008).

A later analysis of the same data showed that, although increasing productivity through predator control increased population viability, the long term prospect was still negative. That seems to be because the growth rate of the whio populations modelled was largely driven by the survival of the breeding adults. Populations with no predator control and low productivity will rapidly decline to extinction (Whitehead et al., 2010).

In Kahurangi National Park, riverside trapping operation combined with pulsed aerial 1080 operations in adjacent forests. Adult female whio survival, probability of breeding, nesting success, and duckling survival were monitored for 7 years through two full beech mast events. Nesting success was highest in the season following a 1080 drop, and lowest at its edges. The combination of trapping and 1080 helped to protect this whio population, especially those at the centre of a large treatment block (Steffens et al., 2022).

Fig. 14.6 A family of whio (blue ducks) on the Mangatepopo River, Ruapehu, November 2005. (Photo by Tyrone Smith: Crown copyright https://creativecommons.org/licenses/by/4.0/)

The survival of whio ducklings is also badly affected when rivers rise in spate, because they can easily be separated from their parents, with fatal consequences. Climate models predict increasing frequency of extreme rainfall events. In two river valleys in northern Te Urewera National Park, one trapped and the other not, data from 13 years of monitoring of birds and river flows were used to construct a population viability model. Craig Simpkins and team concluded that the combination of losses due to predation and to flooding events increased the risk of local extinction of whio by 69%. Trapping can, at the moment, maintain the population, but if it stopped, or if the number of severe floods doubled, whio would rapidly disappear from those catchments (Simpkins et al., 2015).

On the other hand, the same treatments that remove stoats from the river banks also kill rats, and rats roam the river banks in far greater numbers than do stoats. Monitoring results, fledging counts, and the models based on them, cannot distinguish between their effects. Flooding events independent of stoat activities further accelerate the risks.

14.2.7 Kākā

Estimates of the survival and nesting success of kākā (*Nestor meridionalis*), a large endemic and colourful cavity-nesting forest parrot, confirm the serious effect of

Fig. 14.7 The kākā is one of a handful of endemic hole-nesting species that has most clearly benefitted from the absence of stoats on Maungatutari Sanctuary Mountain. (Photo by Author)

predation by stoats on unprotected eggs, chicks, and adult females. Fenced sanctuaries can quickly reverse those processes (Fig. 14.7). Kākā are therefore key indicator species for monitoring programmes designed to suppress stoat populations, because no other invasive predators affect kākā as strongly as stoats do.

Sitting females (only the females brood) and fledglings are especially vulnerable, because they nest in tree holes with only one entrance, and have long periods of incubation (24 days) and fledging (70 days). Peter Wilson led a long term study of one unprotected forest area in the Nelson Lakes region, monitored for 11 years, through six beechmast cycles. Only four young kākā survived to independence (two of them from a nest protected from climbing predators) out of 20 breeding attempts. Nine young and four adult females were killed on their nests. Kākā are too big for ship rats to tackle directly, and they attempt to breed only when food is abundant, therefore much more often in beechmast years when stoats are more numerous . Differential mortality of sitting female kākā due to stoats distorts the sex ratio of breeding adults, increases competition among males for mates, and accelerates the population decline (Wilson et al., 1998).

Kākā are common on offshore islands where stoats are absent, which suggests that stoats could eventually extinguish all unprotected mainland populations of kākā.

In the Eglinton Valley, stoat trapping over 13,000 ha with 193 Fenn trap sites set to protect kākā and mohua began in late 1997, when 35 stoats were caught. Over the next two summers (including the beech seedfall year of 1999), 217 and 274 stoats were caught. Kākā nests were monitored over two seasons (1998/99, 1999/2000).

Of 25 nests, 80% successfully fledged young, and all 13 radio-tagged chicks survived. By contrast, in unprotected beech forests at Lake Rotoiti, where kākā were monitored in the absence of predator control, only 10% of nests produced young, and only 15 of 30 Rotoiti fledglings survived to a year old. A typical post-seedfall irruption of rats accompanied this study (Dilks et al., 2003).

A review by Ron Moorhouse et al. (2003) of three studies of kākā breeding in areas with predator control and three without found that, at protected nests, the number of chicks fledged was 2.5, 2.3, and 2.2, 15% of 70 nests failed, and 5% (2/38) radio-tagged breeding females died. The equivalent figures for unprotected nests were 0.25, 0.4 and 1 fledglings, 84% of 43 nesting failures, and 65% (11/17) of females killed (usually by stoats). At all three managed sites, the estimated recruitment of adult females significantly exceeded adult female mortality, but at none of the unmanaged sites. Predator control raised total nesting success from <38% to >80%. Even where stoats are only suppressed to low levels, rather than entirely eradicated, kākā can still stage a dramatic recovery, as observed on Waiheke Island (Sect. 7.4). The annual 5-min bird counts conducted on Waiheke every November (2020–2024) have found an average annual increase in counts of kākā of 97%, and of smaller native birds of 19% (Holmes, 2025).

It is true that competition among male kākā for the outnumbered adult females can be savage. Possums can compete for nest holes, so kākā increased on Kapiti Island after possums were eradicated. Introduced vespulid wasps compete with kākā for supplies of nutritious honeydew. But kākā are probably the best indicator species for monitoring the success of a stoat control operation because they are not affected by ship rats, ferrets or dogs. The consequences of stoat predation for other sensitive species (kākāriki (*Cyanoramphus* sp), robins/tātāwai (*Petroica longipes, P. australis*), saddlebacks/tīeke (*Philesturnus carunculatus, P. rufusater*)) cannot be distinguished from those of ship rats.

14.2.8 Small Bush Birds

Early ornithologists often commented on the devastating effects of stoats on native birds (O'Regan, 1966) but at that time it was impossible to disentangle the effects on birds of predation versus habitat destruction or prior rat invasions. There was also a general theoretical assumption that, at least in the northern hemisphere, predators take only the old, the weak and the surplus, but there were no hard data to test whether that was true in New Zealand. A 1983 suggestion, that the summer irruptions of stoats described in 1976 and 1979 could be periods of greatly increased predation on birds by rats as much as stoats, fell flat for lack of field data on rats (Sect. 5.4). There were some measurements of the effects of predation on the local productivity of the birds, but few estimates of consequences for bird populations long term.

Stoats are fully capable of climbing rapidly up and down trees and searching the canopy for nests. Bite marks on carcasses or eggs taken by stoats show paired punctures 4.35–9.04 mm apart, exactly matching the placing of the canines in a stoat's jaw (King & Moody, 1982: 72). So small native passerines such as fantails/piwakawaka (*Rhipidura fulginosa*), robins and tomtits/miromiro (*Petroica macrocephala*) (Fig. 14.8) can suffer very high nestling mortality from mustelids as well as rats.

Mustelids are scarcer than rats but can find relatively more nests with chicks than can rats, because the acute directional hearing of mustelids makes it easy for them to find the nests by locating the calling chicks (Sect. 14.2.3). Rats find nests with eggs at random, simply by searching the canopy. In such a dangerous habitat, breeding adults have to re-nest several times in a season. Nevertheless they are remarkably tame and friendly birds, and will often approach human field workers in the expectation that their activities will stir up invertebrates from the leaf litter (Fig. 14.8).

Over seven summers during the 1970s, the dynamics of South Island robins living in a patch of bush near Kaikoura were compared with those of three predator-free offshore islands (Flack & Lloyd, 1978). By contrast with the island birds, which presumably still live in more or less their natural state, the mainland birds studied at Kaikoura had much larger breeding territories (1–5 ha, instead of 0.2–0.6 ha); their 6 month breeding effort started much earlier (July–August) and finished much later (late January) than the 3–4 month season (late August to November) on Outer Chetwood Island; their productivity was higher (clutch size averaging three instead of two, three broods reared per season instead of one, and 3.0 juveniles fledged per pair per season instead of 0.14–1.1); nest failure due to predators averaged 55% per season instead of less than 10%; and annual adult mortality was 23–37% instead of 17%. Despite this enormous extra effort, the mainland birds lived at a much lower density (5–9 ha per pair, instead of 0.4 ha per pair) and not all available breeding habitat was used every season.

Only the female robins brood the nestlings, for about 80% of daylight hours (Powlesland, 1983), so they were more likely to be caught on the nest when a predator found it. The mainland robins survived by making a huge effort to compensate

a b

Fig. 14.8 (**a**) A South Island robin watching for litter invertebrates stirred up by field staff on their trap rounds. (Photo by Author). (**b**) The South Island tomtit is among several species of small birds that still survive in the bush, although much less common than they once were. (Photo by Author)

for their losses, but elsewhere they and many other species could not. It is not hard to see why so many populations of small native bush birds are now becoming severely threatened.

In Nelson Lakes National Park, standard bird counts along an elevational gradient recorded over a span of 30 years documented a steady decline in the abundance of five species of common, widespread native bush birds (bellbird/kōmako (*Anthornis melanura*), rifleman/titipounamu (*Acanthisitta chloris*), grey warbler/riroriro (*Gerygone igata*), tui *Prosthemadera novaseelandiae*) and tomtit) at low but not at high elevations. Three other native species increased in abundance: silvereye/tauhou (*Zosterops lateralis*), yellow-crowned parakeet/kākāriki (*Cyanoramphus auriceps*) and robin, and introduced blackbirds were unaffected. Stoats, rats and, more recently, possums and wasps were considered responsible for these substantial changes in bird community structure observed (Elliott et al., 2010). Similar variable results of a restoration programme were reported from five islands in the eastern Bay of Islands. Recovery of native forests and food webs after pest control favoured long-established native birds at the expense of non-natives or recent immigrants (Ralph et al., 2020). The effects of protecting native birds is best seen in fenced sanctuaries (Sect. 13.2).

At Craigieburn, an isolated patch of mountain forest in Canterbury, a paired-catchment experiment led by Dave Kelly compared the nesting success of bellbirds in response to removal of stoats for 10–12 weeks during the 2000/01 and 2001/02 nesting seasons. Bellbird numbers per 5-min count increased by 79%, and remained higher for 2 years after the end of trapping. In 2000–2001 bellbird nests in the trapped area were four times more likely to succeed than in the non-trapped area (64% compared with 16%) (Kelly et al., 2005).

An extension of that study analysing almost 11,000 5 min bird counts run in 1978–82, 1999–2004, and 2019–2020 found bellbirds continued to increase long term. Stoat traps were set during only 9 of the years studied (1973–2004), but continuously over the last 13 years. Three of nine native bird species declined during the first 25 years, but after the start of sustained trapping in 2007, the counts for 2019–20 found three native bird species increasing in response to consistent removal of stoats (bellbird, rifleman, tomtit), and only one decreasing (brown creeper/pīpipi). The six introduced bird species benefitted from beechmast seeding. It all looks encouraging, but the data set is very complex, and complicated by the arrival of ship rats at Craigieburn since 2010 (Rossignaud et al., 2022).

John Innes'2010 re-analysis of NZ Bird Atlas squares found 18 mainland species at risk of serious decline, of which 9 are endangered or critically threatened. Predator control to prevent further damage to these species is necessary for all 18, but only 7 are getting it. Only 15 species don't need help at the moment, but all common species are at risk of moving into higher threat categories over time (Innes et al., 2010). A massive meta-analysis of the responses of 20 living bird species to various levels of predator control confirmed that most populations do reliably benefit from removal of predators generally, not only stoats (Fea et al., 2020).

On the other hand, long before any mustelids arrived, two species of rats had in turn spread throughout the mainland forests (Polynesian rats *Rattus exulans* in the thirteenth to eighteenth centuries, and Norway rats (*R. norvegicus*) from the 1780s to 1890s). The massive mortality imposed by rats on bird species previously adapted to a live-long, breed-slowly life style had a drastic and unappreciated effect on native bird populations. The agile ship rat (*R. rattus*), by far the worst bird predator of the three, reached the South Island in the 1890s, and at about the same time, Norway rats disappeared as stoats spread west over the main divide from the eastern sheep country (Sect. 1.2).

Ship rats are massively more abundant than stoats. They arrived in the North Island earlier than stoats, and have had well-documented and devastating effects on bird faunas in many places where stoats are still absent, including many valuable offshore islands (Towns, 2023). In a patch of good North Island forest in January, there might be 4–6 ship rats per ha (Hooker & Innes, 1995) or 400–600/km^2, and fewer than 5 stoats/km^2 (up to 10/km^2 after a peak mouse year). The toll does not depend on predatory skill, just *numbers*.

Ship rats and possums are as capable tree climbers as are stoats (Foster et al., 2011), and contribute most to the losses of at least 26 million eggs and chicks every nesting season. According to one famous but unconfirmed back-of-an-envelope calculation, predators kill more eggs, nestlings and their parents in all New Zealand forests every night than the 20,000 that died during the *Rena* oil spill (Hill, 2012). Dietary calculations are greatly influenced by the sample location. One estimate by Chris Jones et al. (2011) suggested that stoats ate too few rats to have any effect on rat populations (Sect. 11.5). That could easily be true in warm, mixed podocarp-broadleaved forest supporting abundant rats and possums, plus some ferrets and cats. Hence, Wendy Ruscoe's large field experiment found that removing stoats alone did not predictably release rat numbers (Ruscoe et al., 2011), in part because other predators remained. By contrast, Kelly Whitau's landscape-scale survey using long term (1998–2016) tracking data in cooler beech forests (Whitau et al., 2023) concluded that effective, consistent targeting of stoats is likely to be followed by higher tracking rates of rats (Sect. 11.5).

John Innes' 2010 landmark survey of predation as a factor limiting New Zealand forest birds (Innes et al., 2010), stating that 17 of 34 threatened or at-risk birds are primarily affected by stoats, is widely cited, but this total does not imply the stoat risk is widespread and general, for three reasons. (1) Five of the 17 species listed are confined to offshore islands, but were included because they would be limited by stoats if ever re-introduced to the mainland, or if a stoat swam to their island (Sect. 7.2). (2) another five of the remaining 12 are separate populations of kiwi (Sect. 14.2.4). (3) the remaining seven come from beech forest populations, which are vulnerable to stoat irruptions after masting events (Chap. 5).

Hence, only two categories of birds are especially vulnerable to stoats: those with large, flightless or long-dependent young (kiwi, kākā, whio), and those confined to beech forests (parakeets/kākāriki, mohua, and brown creeper/pīpipi (*Mohua novaeseelandiae*). Later information adds the alpine specialists (takahē, rock wren) to the list. All the others in a long list are occasionally killed by stoats, but their populations are more at risk for other reasons.

Conservation action does not always produce the intended result. In mixed podocarp-hardwood North Island forest at Mapara in 1989 before a 1080 drop, 71% of stoat guts contained rats, and 6% birds; after the rats had gone, these proportions switched to 17% and 56%. Suddenly deprived of rats, stoats took more birds, the nearest substitutes (Murphy & Bradfield, 1992). This early observation was an unpleasant reminder that human efforts to protect birds can sometimes backfire. A follow-up study at Okarito confirmed that, not surprisingly, stoats ate more rats when there are more rats available to eat, whereas stoats ate more birds as the number of rats available fell (so the two were strongly *negatively* correlated) (Sect. 5.4). These figures are clear evidence of a consistent trend for stoats to switch to stoats eating more birds following successful poison operations against rats, but doesn't mean that stoats are the primary threat to bird populations everywhere.

In North Island mixed forests, 65 nests of New Zealand robins and tomtits were monitored in the 1993–1994 breeding season, of which 72% were lost to ship rats (Brown, 1997). No stoats were seen at the nests during this study, so (assuming that there could have been stoats present) it looks as if stoats were probably not important predators of forest passerines in non-beech forests compared with rats. Any benefits of removing of rats, mice and possums is lost within months, as all three of these pests rapidly recover (van Heezik et al., 2020).

Unexpected results may also follow from island eradications. For example, Colin Miskelly's survey of 21 small Fiordland islands (most <10 ha) compared bush bird count data from both before and after they had been cleared of predators. Ship rats had never established on most of the islands surveyed; Norway rats were eradicated in the 1980s, and stoats have been effectively absent since 2008. South Island robins and kākā nest in tree holes and have significantly increased in numbers, and spread to other islands. But six other species were encountered on significantly *fewer* islands; native pigeon/kereru, tomtits, grey warblers/riroriro, and silvereyes/tauhou, plus the introduced dunnock (*Prunella modularis*) and chaffinch (*Fringilla coelebs*). Direct aggression by robins is the most likely reason why these smaller bird species have declined on (or are excluded from) small islands when robins become abundant (Miskelly et al., 2021).

14.2.9 Ground-Nesting Birds

Stoats frequently carry prey back to their dens, and stoats carrying radio-collars make it easy to find the many scats and food items stashed there. In the Tasman Valley, most of the 219 stoats dens searched contained prey remains, especially lagomorphs and birds (69% each). The remains of birds were larger and easier to identify than from scats, because stoats often left whole discarded heads or wings in or near their dens. Most frequent were the remains of 24 adult wrybill (*Anarhynchus frontalis*) in 8% of dens and 25 adult black-fronted terns/tarapiroe (*Chlidonias albostriatus*) in 5% of dens—in each case this number constituted about 20% of their populations in the area at the time. Banded dotterels/tuturiwhatu (*Charadrius bicinctus*) were found in 15% of dens (Dowding et al., 2015).

The endemic northern New Zealand dotterel/tuturiwhatu (*Charadrius obscurus*) nests in coastal dunes open to the public. Adults are long-lived, but many nesting attempts fail to hatch, even where important nesting sites are roped off and managed by wardens. Predation at dotterel nests on Matakana Island over 6 years was attributed to cats (17 nests), stoats (10) and Norway rats (9), plus black-backed gulls/karoro (*Larus dominicanus*), until they were suppressed. About 20% of the northern subspecies population of NZ dotterels is protected, and stable at present (Dowding & Murphy, 2001). The southern subspecies of NZ dotterels is confined to Stewart Island, where there are no mustelids but many feral cats. On the other hand, habitat destruction and disturbance by people, dogs and vehicles are a constant threat to dotterel nests.

The Hutton's shearwater (*Puffinus huttoni*) has a long-established breeding colony in the alpine zone of the Seaward Kaikōura Range, and during the nesting season, it supplies superabundant food for stoats. Yet, although remains of Hutton's shearwaters were found in 785 of 788 stoat scats collected over three seasons in the Seaward Kaikōura Mountains, the colony is large enough to tolerate predation by stoats at an average rate 0.25% of adults and 12% of chicks per year, and the colony's nesting success is in less danger from stoats than from feral pigs (Cuthbert & Davis, 2002). By contrast, the survival of the kea (*Nestor notabilis*), an endemic alpine parrot, certainly is at risk after a masting year in mountain forests. Kea populations survive because the adults are normally long-lived, but stoats (and cats) can threaten this precious breeding stock (Kemp et al., 2022).

Surviving nesting colonies of yellow-eyed penguins/hoiho (*Megadyptes antipodes*) are few, and often threatened by predators including rats, cats and stoats (Ratz, 2000). One adult Fiordland crested penguin/tawaki (*Eudyptes pachyrhynchus*) was seen making for the scrub, loudly squawking, with a stoat hanging from its neck (Morrison, 1980).

Stoats are agile and excellent swimmers, so it was inevitable that they would invade what remain of New Zealand's once extensive wetlands. Their arrival was first described by Guthrie-Smith (1969). By 1940, he was grieving that 'Where after floods on areas of drying mud there used to be revealed the imprints of the Swamp and Water Crakes, no little footmarks are now visible' (ibid. p. 215). Stoats are well known for active hunting of ground nesting birds along braided rivers (Cruz et al., 2013). Video footage of a stoat destroying eggs and chicks at a banded rail nest, and then sleeping among the remains, was recorded by Parker and Brunton (2004).

On the other hand, stoats were not responsible for some of the worst wildlife disasters in New Zealand history. Norway rats were the first alien predators to reach Fiordland islands such as Anchor, Secretary and Resolution. All the most significant historic losses of previously untouched Fiordland landbirds and seabird colonies were due to them. At the time these voracious predators were distributed around the coasts and offshore islands by sealers and whalers during the nineteenth century, the native avifauna was still more or less intact. The persistent destruction by Norway rats of kākāpō, SI piopio, SI saddleback/tīeke, little spotted kiwi, now all completely gone, and of un numbered seabird nesting colonies, were wiped out and unrecorded, long before stoats arrived (Sect. 7.3). Stoats never reached the last refuges of endemic birds invaded by ship rats on Big South Cape (Bell et al., 2016) or by cats on Stephens (Galbreath & Brown, 2004) Islands.

14.2.10 Bats

The lesser short-tailed bat *Mystacina tuberculata* is the last living representative of the Mystacinidae, an ancient bat family originating in Australia at least 40 million years ago. At least four members of the family had been windblown across the Tasman Sea by about 35 million years ago. Prolonged isolation in New Zealand's forests has made their descendants different from any other bats in the world, endemic at the Family level (O'Donnell et al., 2023). They have become 'walking bats', capable of tucking their wings into protective pockets and foraging on the ground. The forest floor was free from snakes and predatory four-footed mammals, so this was a good hunting strategy that avoids the cost of flying. Of the other three species of mystacinids, one has not been seen for decades and is presumably extinct, two are known only from fossils, and several more remain undescribed. Meanwhile, the Australian lineage, known only from fossils, is now extinct.

Short-tailed bats once numbered in the millions throughout New Zealand, and still survive in thousands, divided into 13 known populations in three recognizable subspecies. The main causes of their decline is deforestation, but where forests still stand, cats, rats and stoats are a constant hazard. The long-term average survival rate of marked short-tailed bats in the Eglinton Valley was 83% until rat numbers increased, then dropped to 40% (O'Donnell et al., 2011). After an effective rat control operation, survival rates recovered to at least 97%. But bats are also vulnerable to secondary poisoning during predator control operations. One such drastic outcome following the use of diphacinone killed more than 100 bats in Pikiariki in 2009 (Dennis & Gartrell, 2015). The only consolation is that large numbers of dead adults and pups were collected, freeze-dried and deposited in the Museum of New Zealand Te Papa Tongarewa.

The long-tailed bat *Chalinolobus tuberculatus* (Fig. 14.9) is the other one of only two surviving native land mammals in New Zealand. It evolved from an Australian species of *Chalinolobus* that arrived unaided sometime during the early Pleistocene (O'Donnell et al., 2023). It has been here long enough to become distinct from its Australian relatives, so is endemic to New Zealand, but only at the species level. Morepork/ruru owls (*Ninox novaeseelandiae*) and the extinct laughing owl/whēkau (*Ninox albifacies*), are, or were, the natural predators of long-tailed bats, but had no effect on bat numbers throughout New Zealand in the 1800s.

Long-tailed bats swarmed everywhere in pre-European times. Maps of the present (2000–2018) distribution data for long-tailed bats maintained by DOC show a wide national spread of remnant records. Bats are surprisingly long-lived for such small animals. In the Eglinton Valley, the oldest known wild female long-tailed bat was 24 years old in 2016, and the oldest male 22 years old. Three colonies in the valley averaged 303 banded individuals in total, plus there were another five

Fig. 14.9 A female long-tailed bat foraging on the forest floor of the Eglinton Valley. New Zealand's native bats can escape the cost of flying by adopting the 'walking bat' pose that evolved in predator-free pre-human times. (Photo by Colin O'Donnell)

colonies in the valley, implying that the local population was more than 800 individuals (O'Donnell & Borkin, 2021).

Before any invasive predators arrived, population sizes would have been significantly greater. Now, survival is lowest during post-seedfall rodent irruption years when stoats and ship rats are abundant (Pryde et al., 2005). Intensive predator control has helped to slow their ongoing decline in some areas, but the foraging ranges of the bats usually exceed the areas of any feasible predator control operation. As for mohua and for the same reason, most damage is done by rats and stoats, both agile enough to climb to nursery roosts, where sleeping adults and pre-volant pups are completely defenceless. Their population decline continues at about 5–9% per year, equating to an 84–96% loss over three generations. Hence the longtailed bat is classed as Nationally Critical (O'Donnell et al., 2023).

14.3 The Case Against Ferrets

Ferrets are semi-domesticated specialist predators of rabbits, which they can hunt above or below ground with equal facility. They were the first mustelids to be transported to New Zealand in the early 1880s, because they were simply assumed to be capable of controlling the plagues of rabbits threatening the main colonial income, wool exports (Sect. 1.2), and because they were the easiest to obtain. Their descendants have mostly remained in open country, and continue to eat rabbits. They do not climb trees, and have never established resident populations on any islands, so they affect a different subset of native fauna compared with agile, forest-dwelling stoats.

14.3.1 Kiwi Adults

Ferrets were still absent from Northland in 1961 (Marshall, 1963), and were slow to spread there until after the late 1970s. Their arrival coincided with a sudden decline in numbers and range of the formerly abundant brown kiwi (Miller & Pierce, 1995). Over 14 years of monitoring at five kiwi sanctuaries established in 2000, ferrets killed 31 adult kiwi (29% of known deaths) (Robertson & de Monchy, 2012). Nevertheless, long-term monitoring of adult kiwi obtained by recording annual standard call indices has been under way at 23 stations in Northland since 1995. By 2002, there was a detectable benefit associated with predator control (Pierce & Westbrooke, 2003).

It could be argued that ferrets are a more significant cause of the long decline in kiwi numbers than are stoats, because ferrets can kill the breeding adults, which take much longer to replace than do chicks. But only one study has provided evidence of how many adult kiwi died from ferret predation as opposed to other causes of death, *combined with* details of the consequences for the productivity of the kiwi population.

Over 20 years in the Tongariro Kiwi Sanctuary (20,000 ha), aerial drops of 1080 were employed every 5 years to clear the forests of rats and mustelids. A total of 85 adult and 57 subadult brown kiwi had been radiotagged over that time, and their fates were known by the time the results of the programme were analysed (Robertson et al., 2019).

The data from four iterations of the five-yearly programme were averaged to give a single figure for the first, second etc. years after the most recent drop. None of the radiotagged kiwi accidentally died from 1080, but 20 were killed by ferrets, in increasing numbers (0, 1, 5, 14) every year since the last 1080 drop. Meanwhile, the survival of chicks to 6 months old decreased in the same order (50%, 29%, 17%, 15%); likewise the survival of adults (98%, 96%, 89%, 87%); likewise and worst, the annual growth of the kiwi population (11.5%, 3.1%, minus 3.6% and minus 5.9%. The combined effect produced an enhanced growth of the kiwi population in the first and second years after the most recent 1080 drop, only to be undone over the subsequent years.

The cherished vision of PFNZ 2050 for restoring national kiwi populations by eradication of stoats has been overtaken by later observations showing that the restoration of kiwi populations does not depend only on protection of chicks from stoats. Rather, the critical factor affecting kiwi population growth is any un-naturally high mortality of adults (Robertson et al., 2011), which can reduce the life expectancy of the normally long-lived breeding stock by two thirds.

Dogs (especially poorly managed hunting dogs) and pigs are at least as damaging to kiwi populations as are ferrets, and have been present in New Zealand for much longer. Predation on adult kiwi by dogs (not included in PFNZ 2050) is a serious risk. One individual dog in Waitangi State Forest systematically hunted and killed more than half of the remaining 900 brown kiwi in less than 3 weeks (Taborsky, 1988). Solutions are now being addressed by various programmes, such as putting dogs through comprehensive kiwi avoidance training. Where dogs are absent or are effectively managed, the main threat to adult kiwi comes from ferrets.

Critics of the use of 1080 (Chap. 8) assert the unacceptable risk to birds of poison operations. But a survey of 748 radio-tracked birds in treated areas compared with 193 in non-treatment areas found that the risk of mortality during a 1080 drop was less than 4% each for kiwi, kākā and kōkako (Veltman & Westbrooke, 2011). Any loss of adult birds is regrettable, but the essential point to understand is that the cost to *individuals* of a successful programme will be far outweighed by the benefit in productivity of the *breeding stock*.

14.3.2 Weka

Weka (*Gallirallus australis*) were an important resource to Māori for food and feathers, and were eagerly hunted by early settlers. Weka were also among the few native predators capable of killing rabbits, but their own food value ensured they could not be given legal protection, as were other rabbit predators at the time (Sect. 1.2). Now, under the Wildlife Act 1953, all remnant weka populations on the mainland are protected. Only the abundant buff weka (*Gallirallus australis hectori*) on the Chatham Islands may be hunted, and then only under a special provision of the Act's third schedule (Fig. 14.10).

Ferrets first arrived in the southern South Island in the late 1870s. Direct contemporary observations record the rapid local extinction of the buff weka on the Benmore Station over the years 1881–1886. Weka frequently occupied rabbit burrows, where ferrets could easily corner them. Canterbury runholder C.J. Tripp reported that in every rabbit hole on his run in which he tried a ferret to turn out the rabbits, out came a weka. Inevitably, the buff weka, the subspecies endemic to the southeastern South Island, was extinct on the mainland by or before 1920. It survives only because in 1905 a group of 12 weka were transferred to the Chatham Islands (Fig. 14.9), which ferrets never reached (King, 2017: 54).

On the other hand, nineteenth-century observations confirmed that weka were very efficient predators of rats, young rabbits and especially mice, perhaps even capable of slowing the rabbit invasion of pastoral lands. William Rees, the Government Sheep Inspector based at Timaru, waged a passionate but ultimately fruitless rear-guard campaign to have the weka protected as a natural enemy of the rabbit, along with ferrets. The Colonial authorities disagreed, so human hunting continued to add to the many weka destroyed by ferrets.

The transformation of vast areas of native tussock grasslands to sheep pasture teeming with rabbits, thistles and Norway rats severely damaged the natural habitat of weka (Rees & Colonial Secretary, 1883–86), and certainly affected their population viability. Charles Douglas (1894) grieved over the arrival of mustelids, fearing that they will finish the birds, and get all the blame, but pointed out that the process of extinction had commenced long before mustelids arrived.

Modern databases and mapping tools now make it possible to compare maps of the distributions of surviving weka populations against the spread of ferrets. Veale (2024) has done that, and concluded that the distribution of ferrets is now largely the opposite of the distribution of weka. By contrast to the Fiordland of Richard Henry's

Fig. 14.10 The buff weka was known to early settlers as the Māori hen, and as good for the pot. A memorial in Dunedin airport remembers the subspecies endemic to Otago, exterminated by settlers and ferrets in the 1880s. (Photo by Author)

time (Henry et al., 2022), there are now no confirmed ferret sightings from the west coast of the South Island, because they don't do well in beech forest, or at altitude. They probably do move westward over the alpine passes in tussocky areas, but then they don't get down through the beech forest to the west coast farmland. But on the eastern side of the South Island, ferrets are abundant enough to prevent the recovery of weka. North Island ferrets probably also do the same, but ferrets are quite patchy there, which might explain the odd few pockets where weka can survive, at least until ferrets arrive.

14.3.3 Lizards

In the drylands of eastern Otago, native skinks have been declining in distribution and numbers since the mid nineteenth century. Rabbits are the primary prey of ferrets and cats, so any drop in local rabbit numbers encourages cats and ferrets to eat more skinks. In addition, grazing by rabbits reduces the vegetation that skinks need for refuges. Predation pressure on skinks increases when rabbits are few, and slackens as rabbits recover, so repeated cycles of intense rabbit control and subsequent recovery are likely to have chronic detrimental effects on skink population viability (Norbury, 2001).

The best evidence for the effects of predation on mainland lizards (and the relative proportions of predator species present) comes from studies demonstrating increased survival and numbers of lizards inside a predator-proof fence, or in the centre of a landscape-scale predator trapping network such as at Macraes Flat,

Otago (Reardon et al., 2012). In that area over 3 years 2006–2008, 2294 hedgehogs were removed, and 635 cats, compared with 423 ferrets, 242 stoats and 42 weasels.

Counts of predators killed in skink habitat often show that other species are much more abundant than any mustelids, and much more dangerous to skinks, especially hedgehogs. Of 158 hedgehogs killed at Macraes Flat in 2000–2001, 33 (21%) had eaten a total of 43 skinks and one gecko, a substantial toll considering the high density of hedgehogs present there (Spitzen-van der Sluijs et al., 2009).

14.3.4 Ground-Nesting Birds

In the Mackenzie Basin, recorded video evidence at 172 nests of ground-nesting birds, demonstrated that ferrets were responsible for 18% of 77 lethal events compared with 43% for cats and 20% for hedgehogs. Floods or trampling by sheep or adult birds accounted for another 4% of nest losses (Sanders & Maloney, 2002). An abrupt decline in the ferrets' primary prey after a local rabbit poisoning operation or an outbreak of the rabbit disease RHD predictably induces ferrets to switch to secondary prey. This sequence of events puts ground-nesting birds and other native fauna at greater risk from predation by ferrets (Norbury & Heyward, 2008).

On the other hand, feral cats and Norway rats are at least as serious a threat to ground-nesting birds as are ferrets. In the Mackenzie Basin following the loss of rabbits due to RHD, 375 cats, 371 ferrets and 86 stoats were killed. All three had eaten birds (47%, 28% and 51% respectively) (Murphy et al., 2004). But the long-term consequences of predation events caused by ferrets versus cats or stoats cannot be distinguished, and the extent of additional mortality due to ferrets versus other mortality agents is usually unknown.

The endemic black stilt/kaki (*Himantopus novaezelandiae*) and its close relative the pied stilt/poaka (*H. himatopus*) are both long-legged wading birds that nest on the ground and feed in the open riverbeds and swamps of the Mackenzie Basin. Both are descended from the same ancestral stock in Australia, but the ancestors of the black stilt arrived so long ago that its living descendants have evolved into a completely different new species. The black stilt has lost several innate antipredator behaviours such as the broken-wing display, and its incubation period has become longer, so extending the most vulnerable period of the brood (Pierce, 1986). By contrast, the pied stilt arrived recently enough to remain identical with its Australian forebears, in appearance and in anti-predator behaviour. Therefore, black stilts are far more vulnerable to predation than pieds, but the decline in their numbers does not depend only on predator control.

The two species of stilts are still interfertile, but pied stilts now hugely outnumber blacks. Black stilts strongly prefer to mate with other blacks, but the shortage of unmated black adults means that some have to accept a mixed partnership. Ongoing hybridisation would be a continuing serious threat to the survival of the black stilt even if every predator in its habitat could be removed (Pierce, 1984).

14.4 The Case Against Weasels

Weasels eat mainly small prey, but little information is available from New Zealand because weasels are rarely caught and samples are small. Recent diet studies and their few predecessors (Strang et al., 2018) found small mammals, reptiles, birds, and invertebrates in various proportions, but these tell us nothing about the weasel's impact on any prey population, unless there are parallel data on the replacement rate of the prey. No such data exist for weasels in New Zealand, but we can learn something from studies of their relatives in England (Chap. 2). Weasels are scarcer than stoats in New Zealand, largely because voles, their primary prey, are absent here (Sect. 1.2).

But when weasels first arrived in the nineteenth century, there were many acceptable ground-feeding substitutes available, including small birds, bats, lizards, frogs, wētā and many large and meaty invertebrates. One man observed a weasel rearing a family of nestlings in a rat hole, and watched her carrying food to them every day, remarking that it was simply 'lizards every time' (Anonymous, 1917). Ever since then, weasels have contributed to the disappearance of small native animals from the mainland.

Weasels and stoats were imported from Britain together, but many more weasels arrived. Their preferred small prey were much more abundant on the ground in New Zealand then than now, and their rapid reproduction rate and short lives (Sect. 2.3) produced a huge but transitory irruption that spread much faster from their liberation point than could stoats. When all the easiest small prey had gone, so did the weasels.

Carterton, in the southern Wairarapa, was chosen for one of the first releases, and from there a sudden tide of weasels spread north to Hawkes Bay. Philosopher-farmer Guthrie-Smith had been observing and minutely describing the abundant wildlife on in his sheep station Tūtira since 1882. He described arrival of weasels there in passionate, vivid terms:

> Weasels, stoats and ferrets were bred and liberated at Carterton, the first-named thriving on a great scale and quickly over-running the countryside. At least before 1901 my then neighbour, John Moore of Rakumoana, noticed weasels on his property. By 1901 they had reached Tangoio, by 1902 a specimen had been seen on Tutira.… the freeing, firstly of rabbits, and secondly of weasels, in the Wairarapa … [was] an outrage, that any individual or local body should have been allowed to attempt to correct a blunder by a crime.
>
> Between [1902] and 1904 they [weasels] had overrun the country between Tutira and the southern edge of the Poverty Bay Flat. Everywhere I heard of them. On every road and new-cut bridle track I met or overtook weasels hurrying northwards.…travelling singly.… For a short period weasels overran like fire the east coast between Tutira and Poverty Bay, and then like fire died out.… Nowadays on Tutira I do not hear from shepherds or fencers of the weasel once in six years. (Guthrie-Smith, 1969: 354–356)
>
> The three years' irruption of weasels through Tutira.…resembled the progress of a comet through the heavens.…the invasion rolled itself up like a scroll; it came and went like a thunderstorm; …then it rained weasels; they poured along the roads. [now] the weasel is a thing of the past, and [has] remained so for many years (ibid.: 389).

Weasels are less likely than stoats to reach unaided any offshore island of wildlife value, but one got to Maud I. in March 2003, and there is a longish record of weasels getting into fenced sanctuaries (Sect. 13.2).

Stoats can suppress weasel populations, and rats suppress mouse populations, so when rat populations are down, mice can rapidly increase, even outside fenced sanctuaries. Therefore, successful removing of stoats and rats risks creating an ideal environment for weasels. Controlled comparisons are few, but 'Remove stoats, get more weasels' is beginning to look like a general rule.

On the other hand, although weasels in general are treated by pest managers as little stoats, they are much smaller and very different in reproductive behaviour. The few data available on their lives in New Zealand generally come incidentally from networks and traps designed for stoats, which are usually too widely placed, and with triggers set to hard, for weasels. All three species of rats have been in New Zealand longer, and done proportionately more damage, than either weasels or stoats.

References

Anonymous. (1917, June 9). The rabbit question. *Southland Times*.

Basse, B., & McLennan, J. (2003). Protected areas for kiwi in mainland forests of New Zealand: How large should they be? *New Zealand Journal of Ecology, 27*, 95–105.

Basse, B., McLennan, J. A., & Wake, G. C. (1999). Analysis of the impact of stoats, *Mustela erminea*, on northern brown kiwi, *Apteryx mantelli*, in New Zealand. *Wildlife Research, 26*, 227–237.

Bell, E. A., Bell, B. D., & Merton, D. V. (2016). The legacy of Big South Cape: Rat irruption to rat eradication. *New Zealand Journal of Ecology, 40*, 212–218.

Blackwell, G. L., Potter, M. A., McLennan, J. A., & Minot, E. O. (2003). The role of predators in ship rat and house mouse population eruptions: Drivers or passengers? *Oikos, 100*(3), 601–613. https://doi.org/10.1034/j.1600-0706.2003.11026.x

Breedt, B., & King, C. M. (2021). Distribution and detectability of mammalian pests in the Waikato Region. *New Zealand Journal of Zoology, 49*, 37–52.

Brown, K. P. (1997). Predation at nests of two New Zealand endemic passerines; Implications for bird community restoration. *Pacific Conservation Biology, 3*, 91–98.

Bunin, J. S., & Jamieson, I. G. (1996). Responses to a model predator of New Zealand's endangered takahe and its closest relative, the Pukeko. *Conservation Biology, 10*, 1463–1466.

Carpenter, J. K., Monks, A., Innes, J., & Griffiths, J. (2022). Pushing the limits: Ship rat (*Rattus rattus*) population dynamics across an elevational gradient in response to mast seeding and supplementary feeding. *Biological Invasions*. https://link.springer.com/article/10.1007/s10530-022-02829-z

Choquenot, D. (2006). Bioeconomic modelling in conservation pest management: Effect of stoat control on extinction risk of an indigenous New Zealand passerine, Mohua ochrocephala. *Conservation Biology, 20*(2), 480–489.

Clout, M. N., & Hay, J. R. (1981). South Island kokako (*Callaeas cinerea cinerea*) in *Nothofagus* forest. *Notornis, 28*, 256–259.

Coté, I. M., & Sutherland, W. J. (1997). The effectiveness of removing predators to protect bird populations. *Conservation Biology, 11*, 395–405.

Cruz, J., Pech, R. P., Seddon, P. J., Cleland, S., Nelson, D., Sanders, M. D., & Maloney, R. F. (2013). Species-specific responses by ground-nesting Charadriiformes to invasive predators and river flows in the braided Tasman River of New Zealand. *Biological Conservation, 167*, 363–370.

Cuthbert, R., & Davis, L. S. (2002). The impact of predation by introduced stoats on Hutton's shearwaters, New Zealand. *Biological Conservation, 108*, 79–92.

Dennis, G. C., & Gartrell, B. D. (2015). Nontarget mortality of New Zealand lesser short-tailed bats (*Mystacina tuberculata*) caused by diphacinone. *Journal of Wildlife Diseases, 51*, 177–186.

Department of Conservation. (2018). *Kiwi recovery plan 2018–2028*. Threatened Species Recovery Plan 64.

Didham, R. K., Tylianakis, J. M., Gemmell, N. J., Rand, T. A., & Ewers, R. M. (2007). Interactive effects of habitat modification and species invasion on native species decline. *Trends in Ecology & Evolution, 22*, 489–496.

Dilks, P. (1999). Recovery of a mohua (*Mohoua ochrocephala*) population following predator control in the Eglinton Valley, Fiordland, New Zealand. *Notornis, 46*(3), 323–332.

Dilks, P., Willans, M., Pryde, M., & Fraser, I. (2003). Large scale stoat control to protect mohua (*Mohoua ochrocephala*) and kaka (*Nestor meridionalis*) in the Eglinton Valley, Fiordland, New Zealand. *New Zealand Journal of Ecology, 27*, 1–9.

Douglas, C. E. (1894). Report on the Westland Alps. *Appendices to the Journal of the House of Representatives, 1894*(C1), 71–74.

Dowding, J. E., & Murphy, E. C. (2001). The impact of predation by introduced mammals on endemic shorebirds in New Zealand: A conservation perspective. *Biological Conservation, 99*, 47–64.

Dowding, J. E., Elliott, M. J., & Murphy, E. C. (2015). Scats and den contents as indicators of the diet of stoats (*Mustela erminea*) in the Tasman Valley, South Canterbury, New Zealand. *New Zealand Journal of Zoology, 42*, 270–282.

Edmonds, H. (2015). *Taxon plan for northern and southern Fiordland tokoeka (Apteryx australis australis): Strategic plan for the recovery of northern and southern Fiordland tokoeka, for the period 2015–2025 and beyond*. Department of Conservation, Te Anau.

Elliott, G. P. (1996a). Mohua and stoats: A population viability analysis. *New Zealand Journal of Zoology, 23*(3), 239–247.

Elliott, G. P. (1996b). Productivity and mortality of mohua (*Mohoua ochrocephala*). *New Zealand Journal of Zoology, 23*(3), 229–237.

Elliott, G., & Kemp, J. (2016). Large-scale pest control in New Zealand beech forests. *Ecological Management & Restoration, 17*, 200–209.

Elliott, G. P., Wilson, P. R., Taylor, R. H., & Beggs, J. R. (2010). Declines in common, widespread native birds in a mature temperate forest. *Biological Conservation, 143*, 2119–2126.

Fea, N., Linklater, W., & Hartley, S. (2020). Responses of New Zealand forest birds to management of introduced mammals. *Conservation Biology, 35*, 35–49.

Flack, J. A. D., & Lloyd, B. D. (1978). The effect of rodents on the breeding success of the South Island robin. In P. R. Dingwall, I. A. E. Atkinson, & C. Hay (Eds.), *The ecology and control of rodents in New Zealand nature reserves* (Information series 4) (pp. 59–66). Department of Lands and Survey.

Foster, S., King, C., Patty, B., & Miller, S. (2011). Tree-climbing capabilities of Norway and ship rats. *New Zealand Journal of Zoology, 38*, 285–296.

Galbreath, R., & Brown, D. (2004). The tale of the lighthouse-keeper's cat: Discovery and extinction of the Stephens Island wren (*Traversia lyalli*). *Notornis, 51*, 193–200.

Gillies, C. A., Leach, M. R., Coad, N. B., Theobald, S. W., Campbell, J., Herbert, T., Graham, P. J., & Pierce, R. J. (2003). Six years of intensive pest mammal control at Trounson Kauri Park, a Department of Conservation "mainland island", June 1996-July 2002. *New Zealand Journal of Zoology, 30*(4), 399–420.

Glen, A. S., Hamilton, T., McKenzie, D., Ruscoe, W. A., & Byrom, A. E. (2012). Kiwi *Apteryx mantelli* population recovery through community-led trapping of invasive non-native mammals in Northland, New Zealand. *Conservation Evidence, 9*, 22–27.

Guthrie-Smith, H. (1969). *Tutira. The story of a New Zealand Sheep Station* (4th ed.). A. H. & A. W. Reed.

Harris, H. A. L., Kelly, D., Innes, J., & Allen, R. B. (2022). Invasive species and thermal squeeze: Distribution of two invasive predators and drivers of ship rat (*Rattus rattus*) invasion in mid-elevation *Fuscospora* forest. *Biological Invasions*. https://doi.org/10.1007/s10530-022-02789-4

Hegg, D., Greaves, G. J., Maxwell, J. M., Mackenzie, D. I., & Jamieson, I. G. (2012). Demography of takahe (*Porphyrio hochstetteri*) in Fiordland: Environmental factors and management affect survival and breeding success. *New Zealand Journal of Ecology, 36*, 75–89.

Henry, R. T., Hill, S., Hill, J., & Jaenecke, V. (2022). *Letters of a naturalist. The field accounts of Richard Henry on Resolution Island*. Putangi Publications.

Hill, G. (2012). Why I enjoyed the Rena disaster. *Forest & Bird, 346*(November), 49.

Holmes, J. (2025). tekorowaiowaiheke.org

Hooker, S., & Innes, J. (1995). Ranging behaviour of forest-dwelling ship rats, *Rattus rattus,* and effects of poisoning with brodifacoum. *New Zealand Journal of Zoology, 22*, 291–304.

Innes, J. (2024). Personal communication.

Innes, J. G., & Hay, R. (1995). The nesting of the North Island kokako (*Callaeas cinerea wilsoni*)—A review of accounts from 1880 to 1989. *Notornis, 42*, 79–93.

Innes, J. G., Hay, R., Flux, I., Bradfield, P., Speed, H., & Jansen, P. (1999). Successful recovery of North Island kokako *Callaeas cinerea wilsoni* populations, by adaptive management. *Biological Conservation, 87*, 201–214.

Innes, J., Nugent, G., Prime, K., & Spurr, E. B. (2004). Responses of kukupa (*Hemiphaga novaeseelandiae*) and other birds to mammal pest control at Motatau, Northland. *New Zealand Journal of Ecology, 28*, 73–81.

Innes, J., Kelly, D., Overton, J. M., & Gillies, C. (2010). Predation and other factors currently limiting New Zealand forest birds. *New Zealand Journal of Ecology, 34*(1), 86–114.

Johnstone, K. C., Garvey, P., & Hickling, G. J. (2023). Invasive mammal control selects for trap-recalcitrant behaviour and personality. *Biological Invasions*. https://link.springer.com/article/10.1007/s10530-023-03191-4

Jones, C., Pech, R., Forrester, G., King, C. M., & Murphy, E. C. (2011). Functional responses of an invasive top predator *Mustela erminea* to invasive meso-predators *Rattus rattus* and *Mus musculus*, in New Zealand forests. *Wildlife Research, 38*(2), 131–140. https://doi.org/10.1071/wr10137

Keedwell, R. J., & Brown, K. P. (2001). Relative abundance of mammalian predators in the upper Waitaki Basin, South Island, New Zealand. *New Zealand Journal of Zoology, 28*(1), 31–38.

Kelly, D., Brindle, C., Ladley, J. J., Robertson, A. W., Maddigan, F. W., Butler, J., Ward-Smith, T., Murphy, D. J., & Sessions, L. A. (2005). Can stoat (Mustela erminea) trapping increase bellbird (*Anthornis melanura*) populations and benefit mistletoe (*Peraxilla tetrapetala*) pollination? *New Zealand Journal of Ecology, 29*, 69–82.

Kemp, J. R., Young, L., Mosen, C., Bolitho, L., Orr-Walker, T., Yockney, I., & Elliott, G. (2022). Irruptive dynamics of invasive carnivores and prey populations, and predator control, affect kea survivorship across the Southern Alps. *New Zealand Journal of Zoology*. https://doi.org/10.1080/03014223.2021.2021249

King, C. M. (1983). The relationships between beech (*Nothofagus* sp) seedfall and populations of mice (*Mus musculus*), and the demographic and dietary responses of stoats (*Mustela erminea*), in three New Zealand forests. *Journal of Animal Ecology, 52*, 141–166.

King, C. M. (2017). Contemporary observations of predation on the buff weka (*Gallirallus australis hectori*) by mustelids in the South Island during the nineteenth century. *Notornis, 64*, 52–55.

King, C. M., & Moody, J. E. (1982). The biology of the stoat (*Mustela erminea*) in the national parks of New Zealand. *New Zealand Journal of Zoology, 9*, 49–144.

King, C. M., Foster, S., & Miller, S. (2011). Invasive European rats in Britain and New Zealand: Same species, different outcomes. *Journal of Zoology, 285*, 172–179.

Lavers, R. B., & Mills, J. A. (1978). Stoat studies in the Murchison Mountains, Fiordland. In Anon (Ed.), *Seminar on the takahe and its habitat* (pp. 222–233). Fiordland National Park Board.

Lee, W. G., & Jamieson, I. G. (Eds.). (2001). *The takahe: Fifty years of conservation management and research*. Otago University Press.

Little, L., King, C. M., & O'Donnell, C. F. J. (2017). Behaviour of stoats (*Mustela erminea*) raiding the nests of rock wrens (*Xenicus gilviventris*) in alpine New Zealand. *Notornis, 64*, 124–135.

Marshall, W. H. (1963). *The ecology of mustelids in New Zealand* (DSIR information series 38). New Zealand Department of Scientific and Industrial Research.

McDonald, R. A., & Murphy, E. C. (2000). A comparison of the management of stoats and weasels in Great Britain and New Zealand. In H. I. Griffiths (Ed.), *Mustelids in a modern world. Management and conservation aspects of small carnivore: Human interactions* (pp. 21–40). Backhuys Publishers.

McDonald, R. A., Webbon, C., & Harris, S. (2000). The diet of stoats (*Mustela erminea*) and weasels (*Mustela nivalis*) in Great Britain. *Journal of Zoology (London), 252*, 363–371. https://zslpublications-onlinelibrary-wiley-com.ezproxy.waikato.ac.nz/doi/epdf/10.1111/j.1469-7998.2000.tb00631.x

McLennan, J. A., Potter, M. A., Robertson, H. A., Wake, G. C., Colbourne, R., Dew, L., Joyce, L., McCann, A. J., Miles, J., Miller, P. J., & Reid, J. (1996). Role of predation in the decline of kiwi, *Apteryx* spp., in New Zealand. *New Zealand Journal of Ecology, 20*, 27–35.

McLennan, J. A., Dew, L., Miles, J., Gillingham, N., & Waiwai, R. (2004). Size matters: Predation risk and juvenile growth in North Island brown kiwi (*Apteryx mantelli*). *New Zealand Journal of Ecology, 28*(2), 241–250.

Michelsen-Heath, S., & Gaze, P. (2007). Changes in abundance and distribution of the rock wren (*Xenicus gilviventris*) in the South Island, New Zealand. *Notornis, 54*, 71–78.

Miller, P. J., & Pierce, R. J. (1995). Distribution and decline of the North Island brown kiwi (*Apteryx australis mantelli*) in Northland. *Notornis, 42*, 203–211.

Miskelly, C. M., Greene, T. C., McMurtrie, P., Morrison, K., Taylor, G. A., Tennyson, A. J. D., & Thomas, B. W. (2021). Species turnover in forest bird communities on Fiordland islands following predator eradications. *New Zealand Journal of Ecology, 45*, 3449.

Moorhouse, R., Greene, T., Dilks, P., Powlesland, R., Moran, L., Taylor, G., Jones, A., Knegtmans, J., Wills, D., Pryde, M., Fraser, I., August, A., & August, C. (2003). Control of introduced mammalian predators improves kaka *Nestor meridionalis* breeding success: Reversing the decline of a threatened New Zealand parrot. *Biological Conservation, 110*, 33–44.

Morrison, K. (1980). Bird and stoat encounters in Fiordland. *Notornis, 27*, 324.

Murphy, E., & Bradfield, P. (1992). Change in diet of stoats following poisoning of rats in a New Zealand forest. *New Zealand Journal of Ecology, 16*, 137–140.

Murphy, E. C., Keedwell, R. J., Brown, K. P., & Westbrooke, I. (2004). Diet of mammalian predators in braided river beds in the central South Island, New Zealand. *Wildlife Research, 31*(6), 631–638. https://doi.org/10.1071/wr03033

Murphy, E., Maddigan, F., Edwards, B., & Clapperton, K. (2008). Diet of stoats at Okarito Kiwi Sanctuary, South Westland, New Zealand. *New Zealand Journal of Ecology, 32*(1), 41–45.

Norbury, G. (2001). Conserving dryland lizards by reducing predator-mediated apparent competition and direct competition with introduced rabbits. *Journal of Applied Ecology, 38*(6), 1350–1361. https://doi.org/10.1046/j.0021-8901.2001.00685.x

Norbury, G., & Heyward, R. (2008). Predictors of clutch predation of a globally significant avifauna in New Zealand's braided river ecosystems. *Animal Conservation, 11*, 17–25.

Norton, D. A., Young, L. M., Byrom, A. E., Clarkson, B. D., Lyver, P. O. B., McGlone, M. S., & Waipara, N. W. (2016). How do we restore New Zealand's biological heritage by 2050? *Ecological Management and Restoration, 17*, 170–179.

O'Donnell, C. F. J. (1996). Predators and the decline of New Zealand forest birds: An introduction to the hole-nesting bird and predator programme. *New Zealand Journal of Zoology, 23*(3), 213–219.

O'Donnell, C. F. J. (2024). Personal communication.

O'Donnell, C. F. J., & Borkin, K. M. (2021). Chalinolobus tuberculatus. In C. M. King & D. M. Forsyth (Eds.), *The handbook of New Zealand mammals* (Families vespertilionidae and mysticinidae) (3rd ed., pp. 95–130). CSIRO Publishing.

O'Donnell, C. F., Edmonds, H., & Hoare, J. M. (2011). Survival of PIT-tagged lesser short-tailed bats (*Mystacina tuberculata*) through a pest control operation using the toxin pindone in bait stations. *New Zealand Journal of Ecology, 35*, 291–295.

O'Donnell, C. F. J., Borkin, K. M., Christie, J., Davidson-Watts, I., Dennis, G., Pryde, M., & Michel, P. (2023). *Conservation status of bats in Aotearoa New Zealand, 2022*. Department of Conservation.

O'Regan, R. (1966, November). The effect of stoats on native birds in Westland. *Forest and Bird*, pp. 4–10.

Parker, K. A., & Brunton, D. H. (2004). Predation of banded rail (*Rallus philippensis*) nests in a saltmarsh habitat. *Notornis, 51*, 240–241.

Peck, A. (2024, February 8). Kiwi in the capital. *Forest and Bird*.

Pierce, R. J. (1984). Plumage, morphology and hybridisation of New Zealand stilts *Himantopus* spp. *Notornis, 31*, 106–130.

Pierce, R. J. (1986). Differences in susceptibility to predation during nesting between pied and black stilts (*Himantopus* spp.). *Auk, 103*, 273–280.

Pierce, R. J., & Westbrooke, I. M. (2003). Call count responses of North Island brown kiwi to different levels of predator control in Northland, New Zealand. *Biological Conservation, 109*, 175–180.

Powlesland, R. G. (1983). Breeding and mortality of the South Island robin in Kowhai Bush, Kaikoura. *Notornis, 30*, 265–282.

Pryde, M. A., O'Donnell, C. F. J., & Barker, R. J. (2005). Factors influencing survival and long-term population viability of New Zealand long-tailed bats (*Chalinolobus tuberculatus*): Implications for conservation. *Biological Conservation, 126*, 175–185.

Ralph, C. J., Ralph, C. P., & Long, L. D. L. (2020). Towards the reestablishment of community equilibrium of native and non-native landbird species in response to pest control on islands in the Eastern Bay of Islands, New Zealand. *Notornis, 67*(2), 437–450.

Ratz, H. (2000). Movements by stoats (*Mustela erminea*) and ferrets (*M-furo*) through rank grass of yellow-eyed penguin (*Megadyptes antipodes*) breeding areas. *New Zealand Journal of Zoology, 27*(1), 57–69.

Reardon, J. T., Whitmore, N., Holmes, K. M., Judd, L. M., Hutcheon, A. D., Norbury, G., & Mackenzie, D. I. (2012). Predator control allows critically endangered lizards to recover on mainland New Zealand. *New Zealand Journal of Ecology, 36*, 141–150.

Rees, W., & Colonial Secretary. (1883–86). *Re protection of the weka as a natural enemy of the rabbit*. Archives New Zealand R17620862.

Robertson, H. A., & de Monchy, P. J. M. (2012). Varied success from the landscape-scale management of kiwi *Apteryx* spp. in five sanctuaries in New Zealand. *Bird Conservation International, 22*(4), 429–444. https://doi.org/10.1017/s0959270912000044

Robertson, H. A., Colbourne, R. M., Graham, P. J., Miller, P. J., & Pierce, R. J. (2011). Experimental management of Brown Kiwi *Apteryx mantelli* in central Northland, New Zealand. *Bird Conservation International, 21*(2), 207–220. https://doi.org/10.1017/s0959270910000444

Robertson, H. A., Craig, E., Gardiner, C., & Graham, P. J. (2016). Short pulse of 1080 improves the survival of brown kiwi chicks in an area subjected to long-term stoat trapping. *New Zealand Journal of Zoology, 43*, 351–362.

Robertson, H. A., Guillotel, J., Lawson, T. O., & Sutton, N. (2019). Landscape-scale applications of 1080 pesticide benefit North Island brown kiwi (*Apteryx mantelli*) and New Zealand fantail (*Rhipidura fuliginosa*) in Tongariro Forest, New Zealand. *Notornis, 66*, 1–15.

Rossignaud, L., Kelly, D., Spurr, E., Flaspohler, D. J., Allen, R. B., & Brockerhoff, E. G. (2022). Trends in bird counts 1978–2020 in a New Zealand *Nothofagus* forest with variable control of mammalian predators. *Avian Conservation and Ecology, 17*(2), 4.

Ruscoe, W. A., Ramsey, D. S. L., Pech, R. P., Sweetapple, P. J., Yockney, I., Barron, M. C., Perry, M., Nugent, G., Carran, R., Warne, R., Brausch, C., & Duncan, R. P. (2011). Unexpected consequences of control: Competitive vs. predator release in a four-species assemblage of invasive mammals. *Ecology Letters, 14*(10), 1035–1042. https://doi.org/10.1111/j.1461-0248.2011.01673.x

Sanders, M. D., & Maloney, R. F. (2002). Causes of mortality at nests of ground-nesting birds in the Upper Waitaki Basin, South Island, New Zealand: A 5-year video study. *Biological Conservation, 106*, 225–236.

Simpkins, C., Perry, G. L. W., Glaser, A., Allerby, T., & Dennis, T. E. (2015). Effects of predation by introduced mammals and mortality due to severe floods on population viability of the endangered blue duck (*Hymenolaimus malacorhynchos*). *Emu, 115*, 146–157.

Sinclair, A. R. E., Innes, J., & Bradfield, P. (2006). Making endangered species safe: The case of the kokako of North Island, New Zealand. *New Zealand Journal of Ecology, 30*, 121–130.

Smith, D. H. V., Wilson, D. J., Moller, H., Murphy, E. C., & Pickerell, G. (2008). Stoat density, diet and survival compared between alpine grassland and beech forest habitats. *New Zealand Journal of Ecology, 32*, 166–176.

Spitzen-van der Sluijs, A. M., Spitzen, J., Houston, D., & Stumpel, A. H. P. (2009). Skink predation by hedgehogs at Macraes Flat, Otago, New Zealand. *New Zealand Journal of Ecology, 33*, 205–207.

Steffens, K., Malham, J., Davies, R., & Elliott, G. (2022). Testing the effectiveness of integrated pest control at protecting whio (*Hymenolaimus malacorhynchos*) from stoat (*Mustela erminea*) predation in beech forest (Nothofagaceae). *New Zealand Journal of Ecology, 46*(1), 10.20417.

Strang, K., Castro, I., Blunden, G., & Shepherd, L. (2018). The diet of weasels (*Mustela nivalis vulgaris*) from Purerua Peninsula, Bay of Islands, New Zealand. *New Zealand Journal of Zoology, 45*, 83–90.

Taborsky, M. (1988). Kiwis and dog predation: Observations in Waitangi State forest. *Notornis, 35*, 197–202.

Tennyson, A. J. D. (2010). The origin and history of New Zealand's terrestrial vertebrates. *New Zealand Journal of Ecology, 34*, 6–27.

Towns, D. (2023). *Ahuahu. A conservation journey in Aotearoa New Zealand.* Canterbury University Press.

van Heezik, Y., Ray, S. M., Jamieson, I. G., Allen, O., & Schadewinkel, R. B. (2020). Impacts of aerial 1080 predator control on nest success and adult survival of South Island robins. *New Zealand Journal of Ecology, 44*, 3047. https://doi.org/10.20417/nzjecol.44.14

Veale, A., et al. (2013). Using genetic techniques to quantify reinvasion, survival and in situ breeding rates during control operations. *Molecular Ecology, 22*, 5071–5083.

Veale, A. J. (2024). Personal communication.

Veltman, C. J., & Westbrooke, I. M. (2011). Forest bird mortality and baiting practices in New Zealand aerial 1080 operations from 1986 to 2009. *New Zealand Journal of Ecology, 35*, 21–29.

Weston, K. A., O'Donnell, C. J. F., Van Dam-Bates, P., & Monks, J. E. (2018). Control of invasive predators improves breeding success of an endangered alpine passerine. *Ibis, 160*, 892–899.

Whitau, K., Kelly, D., Galloway, T. N. H., MacFarlane, A. E. T., van Vianen, J. C. C. M., Rossignaud, L., & Doherty, K. J. (2023). Effects of altitude, seedfall and control operations on rat abundance in South Island Nothofagus forests 1998–2016. *New Zealand Journal of Ecology, 47*, 3502.

White, P. C. L., & King, C. M. (2006). Predation on native birds in New Zealand beech forests: The role of functional relationships between stoats and rodents. *Ibis, 148*, 765–771.

Whitehead, A. L., Edge, K. A., Smart, A. F., Hill, G. S., & Willans, M. J. (2008). Large scale predator control improves the productivity of a rare New Zealand riverine duck. *Biological Conservation, 141*(11), 2784–2794. https://doi.org/10.1016/j.biocon.2008.08.013

Whitehead, A. L., Elliott, G. P., & McIntosh, A. R. (2010). Large-scale predator control increases population viability of a rare New Zealand riverine duck. *Austral Ecology, 35*(7), 722–730. https://doi.org/10.1111/j.1442-9993.2009.02079.x

Wilson, P. R., Karl, B. J., Toft, R. J., Beggs, J. R., & Taylor, R. H. (1998). The role of introduced predators and competitors in the decline of kaka (*Nestor meridionalis*) populations in New Zealand. *Biological Conservation, 83*, 175–185.

Worthy, T. H., Hand, S. J., Nguyen, J. M. T., Tennyson, A. J. D., Worthy, J. P., Scofield, R. P., & Archer, M. (2010). Biogeographical and phylogenetic implications of an early Miocene wren (Aves: Passeriformes: Acanthisittidae) from New Zealand. *Journal of Vertebrate Paleontology, 30*, 479–498.

Part VI

Part VI Review and Conclusions

Reviewing the Evidence

15

Abstract

Looking back over the long history of our struggle to protect New Zealand's native fauna from stoats and other introduced predators, we can see that determination, new technology and passionate advocacy have won some important battles, but we have not yet won the war. It helps to summarise what we know now, in order to make important contributions to the ongoing debate on the prospects for future management. But the one solid conclusion already available is that native fauna can recover given enough good habitat, reliable food resources and protection from predators. Highly motivated community trapping operations and fenced sanctuaries can supply these; further progress will depend on as-yet unknown future technology.

Keywords

Prospects for management stoats · Predator suppression · Predator Free New Zealand 2050 · Lessons from history

15.1 What We Know Now

When English weasels and stoats were brought to New Zealand, they became transformed from treasured natives to serious pests. Ferrets have a long history of domestication, so are not English natives in the same sense, but are equally unwelcome. The damage done by stoats and ferrets (and, to a lesser extent, weasels) to the vulnerable fauna they encountered is now well documented, although they are often also blamed for losses actually due to rats. Norway rats were common on the ground in native forest from the late 1770s until the late 1890s, but are now replaced by smaller, semi-arboreal ship rats (King et al., 2011).

C. M. King, *Stoat in the Dock*, Fascinating Life Sciences,
https://doi.org/10.1007/978-981-96-4562-6_15

All three mustelid species arrived together in the 1880s. The remains of ancient New Zealand then still supported many large flightless birds as well as Norway rats, although not for much longer. Ferrets and weasels have no delayed implantation, so, given abundant food supplies, both could multiply and spread very rapidly, with catastrophic effects (Sects. 14.3 and 14.4). The arrival of ferrets and stoats (often confused with 'weasels') in Westland and Fiordland in the 1890s was witnessed by Charles Douglas (1894), A. P. Harper (1896) and Richard Henry (2022). Likewise, the arrival of weasels in Hawke's Bay in 1902 was witnessed by Herbert GuthrieSmith (1969). Now, ferrets and weasels have gone from all those areas, replaced by stoats, which can climb better than ferrets and can take larger prey better than can weasels.

The difficulty of deciding what to do about them is complicated by the great many other environmental changes that have also contributed to the catastrophic decline in New Zealand's biodiversity. Ultimately, it seems that Nature is on their side.

For some relatively well-studied threatened forest bird species reviewed in Sect. 14.2, the case against stoats is strong. Ongoing and expensive targeted stoat control to protect them is indeed urgent and fully justifiable. The parallel case against ferrets that damage kiwi adults, weka and lizards is also convincing. For most other native species, there is often plenty of contributory evidence against stoats or ferrets, but it is seldom distinguishable from the other multiple and interacting causes of biodiversity decline, mostly due to habitat loss and to ship rats.

15.2 Prospects for Management

Adaptive evolution is the result of repeated conversations between uncountable generations of animals and their environment. Therefore, an appreciation of the evolutionary history of mustelids helps to understand historic failures. Broadly, our urgent attempts to manage these animals by conventional methods amounts to working against nature. There is no great consolation in admitting that, with hindsight, we see that small mustelids were the very worst possible natural enemies to choose for an irreversible experiment in biological control in nineteenth-century New Zealand. The only worst ones, which fortunately were not an option at the time, would have been American mink.

Intelligent pest management requires that the natural control of animal numbers must be known before rational strategies for artificial management can be designed. All three small mustelids are active, intelligent, widespread, and resilient to periodic fluctuations in prey resources. Lethal control to reduce mustelid numbers from the top down certainly works well locally, as we can see from the encouraging reports from the Predator Free New Zealand 2050 (Chap. 13). But, outside fenced sanctuaries and safe offshore islands, unstoppable reinvasions remind us that their populations are actually controlled by food supplies, from the bottom up. Their 'breed early, live fast, die young' life-history strategy makes the small mustelids extraordinarily resistant to unstable food resources and human management.

This review therefore can only offer some suggestions to explain why mustelids, especially stoats, are so difficult to manage.

1. Effective top-down control measure(s) must exceed the natural mortality of the target population, by adding to, rather than replacing, the normal losses of each annual crop of young by 50–70% a year, rising to 90% in rodent crash years.
2. Young female stoats are fertilised before they leave the nest, so do not need to find a mate to establish a new population, say on a previously pest-free island. Nearly all female stoats of all ages are carrying a potential new litter for 9 months a year. Delayed implantation provides a long opportunity for human managers to remove fertilised females before they have given birth, or to sterilise them in the field if that becomes feasible, but both are at present impossible.
3. When food is super-abundant, female stoats can achieve high natural productivity, increasing from zero to 13 or more independent young per female. Unlike stoats, weasels can breed twice a year, but only when food is super-abundant. Both species evolved to respond to intermittent resource pulsing in the Arctic, so were already pre-adapted to equivalent short-term pulsing by masting cycles of rodents in New Zealand beech forests. Some adult stoats can survive a rodent crash on alternative resources and are capable of widespread dispersal in search of them. Hence, only a few stoats and no weasels live until the next mast year.
4. The most significant factor controlling stoat populations is juvenile mortality, which is closely linked to the effects of prey numbers on the fertility and overwinter survival of fertilised females of all ages, and on the mortality rate of nestlings. After a massive seedfall in southern beech forests, survival among the single annual litter is high, so the huge numbers of stoats that emerge are all young of the year. Starvation cuts down most of them within a few months of independence, and trapping removes many of that annual cohort that would not have survived to breed anyway. In other years, the same processes have less drastic effects.
5. Traps set widely spaced and infrequently checked (to save labour) catch mostly surplus males, and miss many potential captures of fertilised females. Removing only males will have no effect on the numbers of young of the year, or on the next year's population.
6. Weasels were imported to New Zealand in greater numbers than stoats, and are disadvantaged by the absence of voles, but where they can find the right conditions, they can multiply much faster than can stoats or ferrets.

The main problem for the Predator Free New Zealand 2050 vision is that the introduced mustelids are members of complex animal communities with multiple cross-connections to other predators and their prey, all influenced by habitat and seasonal conditions. Stoats and ferrets (and cats) tend to favour rats, possums and rabbits as primary prey wherever these are available, because they get more energy from those mammalian prey than from birds. But all the invasive predators are flexible and opportunist hunters, so when pest managers achieve effective removal of rats and rabbits, predators have to find alternative supplies, and that often means greater dangers for birds. The only guaranteed way of avoiding that outcome is to remove all introduced mammals together.

That is, of course, being done in pest-fenced sanctuaries and on remote islands. The irony is that while huge expenditure is lavished on PFNZ 2050 projects despite their lack of defined biodiversity targets or permanent results, the Department of Conservation spends almost nothing on pest-fenced sanctuaries, which have provided proven long-term benefits for many important species. So New Zealand authorities are not proceeding with the only policy on the mainland that we know does actually work (Innes, 2024).

Three species of rats arrived in New Zealand centuries earlier than mustelids (King, 2019), they attain densities orders of magnitude higher than mustelids can, and have been carried to islands far out of reach of any mustelid. If the same arguments were to be assembled against rats, the evidence for their impact could be gathered from many, many more native species, and many more places that no mustelids ever reached, and in total would be over-whelming.

15.3 Predator Free New Zealand 2050

Of course, bringing mustelids to New Zealand was a disastrous human misjudgment, but the question we have to grapple with now is how far a historic mistake of such magnitude can be corrected by contemporary policy. The unprecedented public support for the PFNZ 2050 programme is enormously inspiring, and is achieving some near-magical diversity gains at least locally and temporarily (Department of Conservation, 2021). These great results are due in part to increasing expertise in lethal technology, yet as the geographic scale of operations increase, reinvasion become more and more unpreventable. So removing predators from important places is easier than keeping them out. Any extension of PFNZ's vision of eradication to the national scale seems out of reach for now, but even partial suppression helps.

Having said that, we should remember that PFNZ's ultimate vision cannot be achieved all at once. It is best seen as **a game of two halves**—what we can do now, and what we hope to be able to do in future (King, 2023). Few people know more about that dilemma from the inside than John Innes. As he says, those us alive now can only focus on the plight we see and the cause now present, and stoats are irrefutably currently on the wanted list. We don't know much, even now, but both theory and practical demonstrations show that when we get rid of all pest mammals, birds thrive. More research is always good, but we can still advance without it until new tools arrive (Innes, 2024).

15.4 Conclusions

More than a hundred years ago, G. M. Thomson's classic work, *The Naturalisation of Animals & Plants in New Zealand* (Thomson, 1922) made the first attempt to summarise what was then known about stoats alongside all the other flora and fauna introduced into New Zealand by Europeans. Forty years ago, I tried to bring the story up to date (King, 1984). Understandably, both books are now of historic interest only. Both worked in ignorance of what is known now, so could not avoid errors of fact, some of which this present book aims to correct (even though history will probably expose a new crop of errors here too). Both previous analyses included errors of judgement that stand as cautionary tales for contemporary management trying to predict wise policies for the future. So much has been learned that 5 years ago an update on the update was needed (King, 2019). The next update will be needed even sooner. That's the path of progress.

New Zealand is better placed than most other countries to learn from its history of invasive predators and of the policies that have been evolved to manage them. The nineteenth-century runholders who imported the mustelids in the first place thought they had found the correct and logical solution to their problem, but they were wrong. If there really is any party to blame for our present situation, perhaps our regrets should aim at their pre-scientific ideas about the balance of nature, rather than at the animals they enlisted to their mistaken cause. Now we are better at asking more critical questions, and at finding the expertise to answer them, but we should not forget that we are not immune to the risk of repeating the mistakes of the past. Our best options, as Albert Einstein so rightly said, are to learn from yesterday, live for today, and hope for tomorrow.

15.5 Tailpiece

The old English nursery rhyme *This is the House that Jack Built* was a memorable cumulative poem designed to explain to children how one thing leads to another, with unforeseen and possibly unpleasant consequences. Its starts with:This is the house that Jack built...This is the malt that lay in the house that Jack built...This is the rat that ate the malt...That lay in the house that Jack built...This is the cat That killed the rat that ate the malt ...and so on.

A New Zealand version relevant to this book might run as follows:This is the house that Jack built.

These are the paddocks, broad and wide
That cover the land on every side
All round the house that Jack built.
These are the grasses, brown and green
That grow on the paddocks rich and wide
That cover the land on every side
All round the house that Jack built.
These are the livestock, woolly and clean

That graze on the grasses, brown and green
That grow on the paddocks rich and wide
That cover the land on every side
All round the house that Jack built.
These are the rabbits, brought in for sport
That eat much more of the grass than they ought
That is needed by livestock, hungry and lean
That search for the grasses, brown and green
Clinging to paddocks muddy and wide
That cover the land on every side
All round the house that Jack built.
These are the imported ferrets and stoats
That kill rabbits cheaper than poisoned oats
But fatally damage the pastoral scene
Where starving sheep, and grass once green
Both wither on paddocks muddy and wide
That cover the land on every side
All round the house that Jack built.
These are the kiwi and native-born friends
That don't have the caution on which life depends
So fall to the imported ferrets and stoats
That kill rabbits cheaper than poisoned oats
And fatally damage the pastoral scene
Where starving sheep, and grass once green
Are scarce on the paddocks bare and wide
That cling to the land on every side
All round the house that Jack built.
These are the tussocks and forests, all gone
That once thronged with kiwi, and rang with bird song
That can't live on paddocks, bare and wide
That scour off the land on every side
All round the house that Jack built.
These are the trappers, hairy and tough
Along with their mates, they've all had enough;
They bring all their various weapons of war -
The traps and the poisons and tunnels and more -
To rescue the wildlife and put right the wrong
That that silenced the kiwi and forest bird song.
They aim to repair the paddocks, still wide,

And restore the forests on every side
All round the house that Jack built.

References

Department of Conservation. (2021). *Predator free 2050 5-year progress report*. Department of Conservation.
Douglas, C. E. (1894). On the Westland Alps. *Appendices to the Journal of the New Zealand House of Representatives, C1*(30), 71–75.
Guthrie-Smith, H. (1969). *Tutira. The story of a New Zealand Sheep Station* (4th ed.). A. H. & A. W. Reed.
Harper, A. P. (1896). *Pioneer work in the Alps of New Zealand*. T. Fisher & Unwin.
Henry, R. T., Hill, S., Hill, J., & Jaenecke, V. (2022). *Letters of a naturalist. The field accounts of Richard Henry on Resolution Island*. Putangi Publications.
Innes, J. (2024). Personal communication.
King, C. (1984). *Immigrant killers. Introduced predators and the conservation of birds in New Zealand*. Oxford University Press.
King, C. M. (2019). *Invasive predators in New Zealand: Disaster on four small paws*. Palgrave Macmillan. https://doi.org/10.1007/978-3-030-32138-3
King, C. M. (2023). Asking the right questions about predator free New Zealand. *New Zealand Journal of Ecology, 47*(1), 3558.
King, C. M., Foster, S., & Miller, S. (2011). Invasive European rats in Britain and New Zealand: Same species, different outcomes. *Journal of Zoology, 285*, 172–179.
Thomson, G. M. (1922). *The naturalisation of animals & plants in New Zealand*. Cambridge University Press.

Index

C. M. King, *Stoat in the Dock*, Fascinating Life Sciences,
https://doi.org/10.1007/978-981-96-4562-6

The manufacturer's authorised representative in the EU is Springer Nature Customer Service Centre GmbH, Europaplatz 3, 69115 Heidelberg, Germany. If you have any concerns regarding our products, please contact ProductSafety@springernature.com

Printed and bound by CPI Group (UK) Ltd, Croydon, CR0 4YY

12/08/2026

02194520-0001